EFFICIENT
BOILER
OPERATIONS
SOURCEBOOK
Second Edition

EFFICIENT
BOILER
OPERATIONS
SOURCEBOOK

Second Edition

F. William Payne

Library of Congress Cataloging-in-Publication Data

Payne, F. William, 1924-
Efficient Boiler Operations Sourcebook.

    Includes index.
    1. Steam-boilers--Efficiency. I. Payne, F. William,
1924-
TJ288.E33   1989   621.1'83   88-82291

ISBN 0-88173-080-7

Published by The Fairmont Press, Inc.
700 Indian Trail
Lilburn, GA 30247

Printed in the United States of America

10 9 8 7 6 5 4 3 2 1

ISBN 0-88173-080-7   FP

ISBN 0-13-245424-6   PH

Distributed by Prentice-Hall, Inc.
A division of Simon & Schuster
Englewood Cliffs, NJ 07632

Prentice-Hall International (UK) Limited, London
Prentice-Hall of Australia Pty. Limited, Sydney
Prentice-Hall Canada Inc., Toronto
Prentice-Hall Hispanoamericana, S.A., Mexico
Prentice-Hall of India Private Limited, New Delhi
Prentice-Hall of Japan, Inc., Tokyo
Simon & Schuster Asia Pte. Ltd., Singapore
Editora Prentice-Hall do Brasil, Ltda., Rio de Janeiro

CONTENTS

INTRODUCTION

The Second Edition of the *Efficient Boiler Operations Sourcebook* contains several expansions on equipment ranges and manufacturers, one minor correction, and a completely new chapter which updates flue gas measurements.

But basically, it remains an applications-oriented book, written to help boiler operators and supervisory personnel improve boiler efficiencies in their plants. Theoretical material has been kept to a minimum. The book concentrates on the three principal fuels—natural gas, oil, and coal.

One set of parameters should be noted. An "industrial boiler," as defined in this book, includes all boilers with 10,000 to 500,000 lb/hr steam flow capacity (10^7 to 5×10^8 heat output capacity) used in either commercial or industrial applications to generate process steam. Utility boilers and marine boilers are excluded.

Three contributors warrant special recognition for their help in developing this book:

Timothy Jones, Product Manager for Thermox Instruments Division of Ametek, Inc., contributed the new chapter (7) updating flue gas measurement techniques.

Richard Thompson, now president of Fossil Energy Research Corporation in Laguna Niguel, California, organized much of the technical data.

Mike Slevin, president of the Energy Technology and Control Corporation in Reston, Virginia, co-authored Chapter 16, "Combustion Control Systems and Instrumentation," and prepared the chapter which follows it, "Boiler O_2 Trim Controls."

Efficient Boiler Operations Sourcebook, Second Edition, includes material originally prepared for the U.S. Department of Energy under D.O.E. contracts C-04-50085 and EC-77-C-01-8675.

F. William Payne

1

Boiler Combustion Fundamentals

Combustion is the complex process of releasing chemically bound heat energy in the fuel through the exothermic reaction of carbon and hydrogen with oxygen to produce carbon dioxide (CO_2) and water vapor (H_2O). In real combustion systems, secondary combustion products such as NOx, SOx, CO and solid particles as well as unburned fuel are released due to the complex make-up of the fuel and incomplete combustion.

While certain gaseous constituents such as NOx and SOx exist only in trace quantities (parts per million, ppm) and are considered important only as air pollutants, other exhaust products such as CO and unburned fuel represent a waste of available heat and are important from an efficiency standpoint.

COMBUSTION AIR

Air consists of 21% oxygen (O_2) and 78% nitrogen (N) and traces of argon and carbon dioxide. For all fuels under ideal burning conditions, there exists a "theoretical amount" of air that will completely burn the fuel with no excess air remaining.

For conventional burners, a quantity of "excess air" above the theoretical amount is required. The quantity of excess air is dependent on several parameters including boiler type, fuel properties and burner characteristics. The quantity of excess air is generally determined by measurements of specific gases (CO_2 and O_2) in the stack and their relation to percent excess air for a particular fuel. These relationships are shown in Figure 1-1.

1

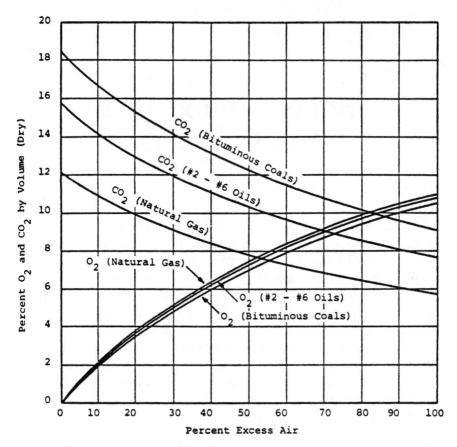

FIGURE 1-1. Relationship between boiler excess air and stack gas
concentrations of excess oxygen (O_2) and carbon dioxide
(CO_2) for typical fuel compositions.

The measurement of excess O_2 is generally preferred over
CO_2 for the following reasons:

- The relation of O_2 to excess air is relatively invarient with
  fuel composition whereas CO_2 relations are fuel dependent.

- CO_2 measurements require more precision than excess O_2
  measures to obtain the same accuracy.

- Excess O_2 is more associated with excess air, i.e., as excess
  air goes to zero, excess O_2 follows.

- Excess O_2 instrumentation is generally less expensive and more reliable.

Stack gas excess air need not reflect combustion conditions at the burners due to air or fuel maldistribution in multiburner systems or air introduction at other portions of the unit.

FUEL CHARACTERISTICS

There are significant differences between the firing system and burning characteristics of the conventional fuels currently in use. Natural gas requires little fuel preparation, mixes readily with the combustion air supply and burns with a low luminous flame. Its simple handling and firing characteristics, and maintenance characteristics have made natural gas the primary industrial fuel in many sections of the country.

Oil fuels require atomization prior to vaporization and mixing with the combustion air supply. The grade of oil (#2 through #6) determines the extent of pretreatment (heating and screening) to achieve proper conditions at the burner atomizer. Mechanical steam and air atomizer systems are used. Oil burns with a bright, luminous flame.

Coal combustion is the most complex of the conventional fuels. Coal firing can be separated into two broad classes: suspension firing and grate firing. The grate properties of coal significantly influence the burner and furnace design, coal handling and preparation equipment, ash disposal methods and the type of precipitator or dust collector installed.

Majority of smaller units (less than 200,000 lbs per hour) are stoker (or grate) fired. Properties of the coal that influence grate design and bed burning include coal fineness, moisture and friability. Larger units utilize suspension firing of pulverized coal with coal grindability and moisture content as important indicators of fuel-bed clinkering and furnace wall slagging.

Fuel analyses of various conventional industrial fuels are given in Table 1-1. These will be used in later efficiency calculations.

BOILER CONFIGURATIONS AND COMPONENTS

Industrial boiler designs are influenced by fuel characteristics and firing method, steam demand, steam pressures, firing characteristics and the individual manufacturers. Industrial boilers can be classified as either firetube or watertube indicating the relative position of the hot combustion gases with respect to the fluid being heated.

Firetube Boilers

Firetube units pass the hot products of combustion through tubes submerged in the boiler water. A typical firetube arrangement is illustrated in Figure 1-2.

Conventional units generally employ from 2 to 4 passes as shown in Figure 1-3 to increase the surface area exposed to the hot gases and thereby increase efficiency. Multiple passes, however, require greater fan power, increased boiler complexity and larger shell dimensions.

Maximum capacity of firetube units has been extended to 69,000 lbs of steam per hour (2,000 boiler hp) with operating pressures up to 300 psig design pressure.

Advantages of firetube units include:
— ability to meet wide and sudden load fluctuations with only slight pressure changes
— low initial costs and maintenance
— simple foundation and installation procedures.

Watertube Boilers

Watertube units circulate the boiler water inside the tubes and the flue gases outside. Typical boiler configurations and general flue gas flow patterns through these units are given in Figures 1-4 and 1-5.

Water circulation is generally provided by the density variation between cold feed water and the hot water/steam mixture in the riser as illustrated in Figure 1-6.

TABLE 1-1. Fuel Analyses

Fuel (Source)	% by Wt As Fired Ult. Analys.							Heating Value Btu/lb	Theoretical Air lb/10,000 Btu
	Carbon	Hydrogen	Oxygen	Nitrogen	Sulfur	Ash	Moisture		
Bituminous Coal									
(Western Kentucky)	71.4	5.0	7.8	1.3	2.8	7.3	4.5	12,975	7.51
(West Virginia)	76.2	4.7	3.8	1.5	1.2	9.0	3.0	13,550	7.58
Subbituminous Coal									
(Wyoming)	56.8	4.1	11.9	0.9	0.8	3.9	21.5	9,901	7.56
(Colorado)	57.6	3.2	11.2	1.2	0.6	5.4	20.8	9,670	7.53
#2 Oil	87.0	11.9	0.6	—	0.5	—	—	19,410	7.27
#6 Oil	86.6	10.8	0.7	1.5	—	—	—	18,560	7.40
Natural Gas (So. Calif.)	74.7	23.3	1.2	0.8	—	—	—	22,904	7.18
Natural Gas (Pittsburgh)	75.3	23.5	—	1.2	—	—	—	23,170	7.18

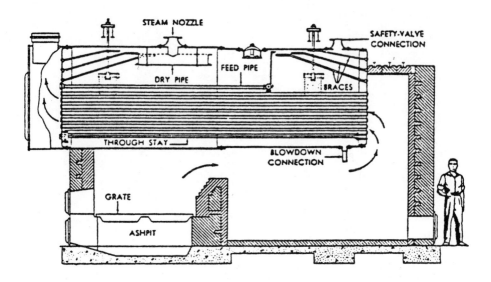

FIGURE 1-2. Sectional sketch of a horizontal-return tubular boiler.

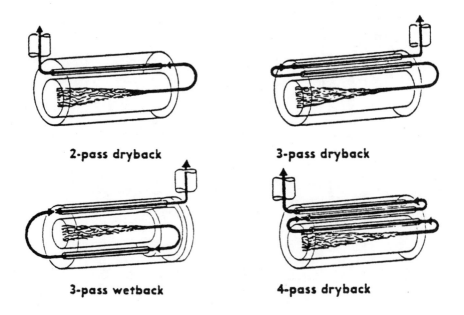

FIGURE 1-3. Typical firetube boiler gas flow patterns.

Watertube boilers may be subclassified into different groups by tube shape, by drum number and location and by capacity. Classifications are also made by tube configuration as illustrated in Figure 1-7. Another important determination is "field" versus "shop" erected units. Many engineers feel that shop assembled boilers can meet closer tolerance than field assembled units and therefore may be more efficient; however, this has not been fully substantiated.

Watertube units range in size from as small as 1000 lbs of steam per hour to the giant utility boilers in the 1000 MW class. The largest industrial boilers are generally taken to be about 500,000 lbs of steam per hour.

Important elements of a steam generator as illustrated in Figure 1-8 include the firing mechanism, the furnace water walls, the superheaters, convective regions, the economizer and air pre-heater and the associated ash and dust collectors.

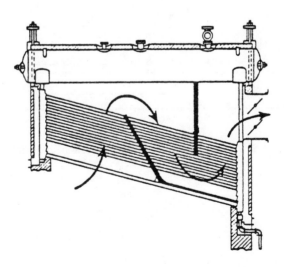

FIGURE 1-4. Small inclined watertube boiler.

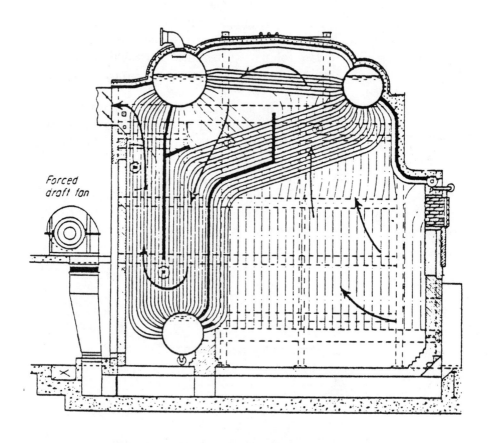

FIGURE 1-5. Bent tube watertube unit typical of industrial applications.

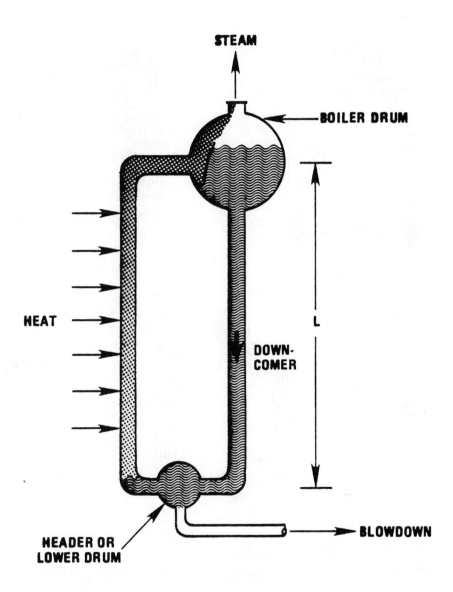

FIGURE 1-6. Water circulation pattern in a watertube boiler.

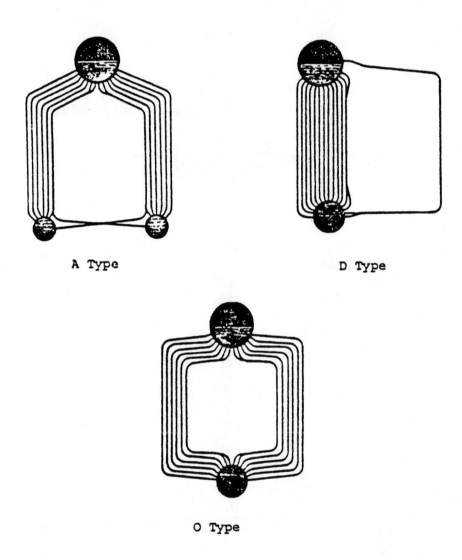

FIGURE 1-7. Classification of watertube boilers by basic tube arrangement.

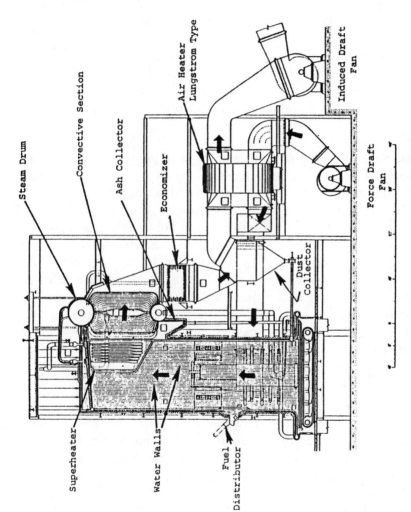

FIGURE 1-8. Layout of the combustion system of an industrial boiler.

FUEL HANDLING AND FIRING SYSTEMS

Gas Fired

Natural gas fuel is the simplest fuel to burn in that it requires little preparation and mixes readily with the combustion air supply.

Industrial boilers generally use low-pressure burners operating at a pressure of 1/8 to 4 psi. Gas is generally introduced at the burner through several orifices that generate gaseous jets that mix rapidly with the incoming combustion air supply. There are many designs in use that differ primarily in the orientation of the burner orifices and their locations in the burner housing.

Oil Fired

Oil fuels generally require some type of pretreatment prior to delivery to the burner including the use of strainers to remove solid foreign material and tank and flow line preheaters to assure the proper viscosity. Oil must be atomized prior to vaporization and mixing with the combustion air supply. This generally requires the use of either air, steam or mechanical atomizers.

The oil is introduced into the furnace through a gun fitted with a tip that distributes the oil into a fine spray that allows mixing between the oil droplets and the combustion air supply. Oil cups that spin the oil into a fine mist are also employed on small units. An oil burner may be equipped with diffusers that act as flame holders by inducing strong recirculation patterns near the burner. In some burners, primary air nozzles are employed.

Pulverized Coal Fired

The pulverized system provides four functions: pulverizing, drying, classifying to the required fineness and transporting the coal to the burner's main air stream. The furnace may be designed for dry ash removal in the hopper bottom or for molten ash removal as in a slag tap furnace.

The furnace size is dependent on the burning and ash characteristics of the coal as well as the firing system and type of furnace bottom. The primary objectives are to control furnace ash deposits and provide sufficient cooling of the gases leaving the furnace to reduce the buildup of slag in the convective regions.

Pulverized coal fired systems are generally considered to be economical for units with capacities in excess of 200,000 lbs of steam per hour.

Stoker Fired

Coal stoker units are characterized by bed combustion on the boiler grate with the bulk of the combustion air supplied through the grate.

Several stoker firing methods currently in use on industrial-sized boilers include underfed, overfed and spreader. In underfed and overfed stokers, the coal is transferred directly on to the burning bed. In a spreader stoker the coal is hurled into the furnace when it is partially burned in suspension before lighting on the grate.

Several grate configurations can be used with overfed and spreader stokers including stationary, chain, traveling, dumping and vibrating grates. Each grate configuration has its own requirements as to coal fineness and ash characteristics for optimum operation. Examples of several stoker/grate combinations are given in Figures 1-9, 1-10 and 1-11.

Spreader stoker units have the advantage that they can burn a wide variety of fuels including waste products. Underfed and overfed units have the disadvantage that they are relatively slow to respond to load variations.

Stoker units can be designed for a wide range of capacities from 2,000 to 350,000 lbs of steam per hour.

Spreader stoker units are generally equipped with overfire air jets to induce turbulence for improved mixing and combustible burnout as shown in Figure 1-11. Stoker units are also equipped with ash reinjection systems that allow the ash collected that contains a significant portion of unburned carbon to be reintroduced into the furnace for burning.

COMBUSTION CONTROL SYSTEMS

Combustion controls have two purposes: (1) maintain constant steam conditions under varying loads by adjusting fuel flow,

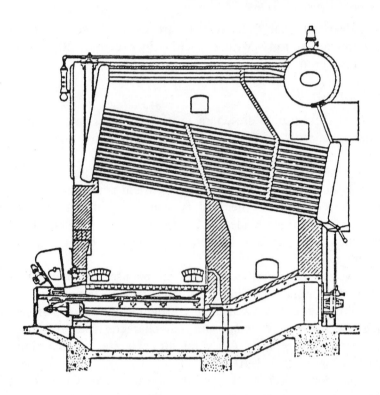

FIGURE 1-9. Single retort stoker (underfed).

and (2) maintain an appropriate combustion air-to-fuel flow. Combustion control systems can be classified as series, parallel and series/parallel as illustrated in Figure 1-12.

In series control, either the fuel or air is monitored and the other is adjusted accordingly. For parallel control systems, changes in steam conditions result in a change in both air and fuel flow. In series/parallel systems, variations in steam pressure affect the rate of fuel input and simultaneously the combustion air flow is controlled by the steam flow.

Combustion controls can be also classified as positioning and metering controls. Positioning controls respond to system demands by moving to a present position. In metering systems, the response is controlled by actual measurements of the fuel and/or air flows.

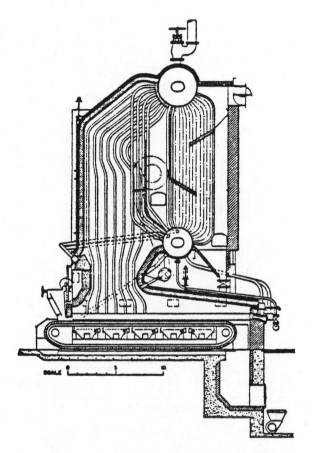

FIGURE 1-10. Steam Generating Unit Equipped with
Traveling-Grate Stoker and Rear-Arch Furnace.

Application

The application and degree of combustion controls varies
with the boiler size and is dictated by system costs.

The parallel positioning jackshaft system illustrated in Figure
1-13 has been extensively applied to industrial boilers based on
minimum system costs. The combustion control responds to
changes in steam pressure and can be controlled by a manual
override. The control linkage and cam positions for the fuel and
air flow are generally calibrated on startup.

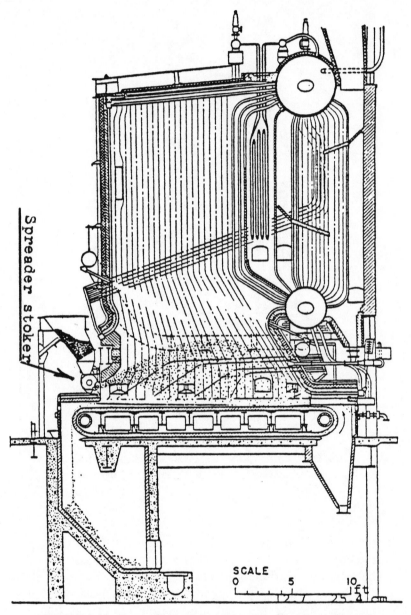

FIGURE 1-11. Steam generating unit equipped with continuous-discharge type of spreader stoker. Rows of overfire air jets are installed in front and rear walls. Cinders are reinjected from boiler hoppers.

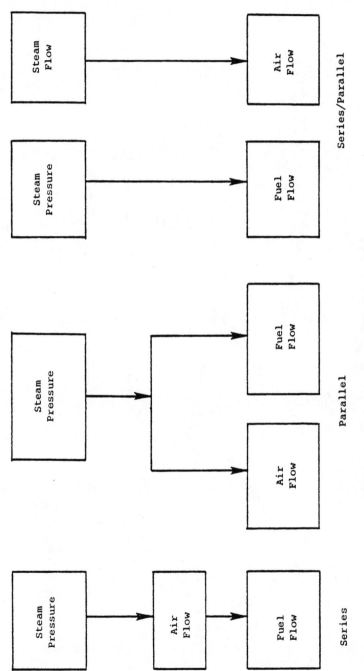

FIGURE 1-12. Basic combustion control systems.

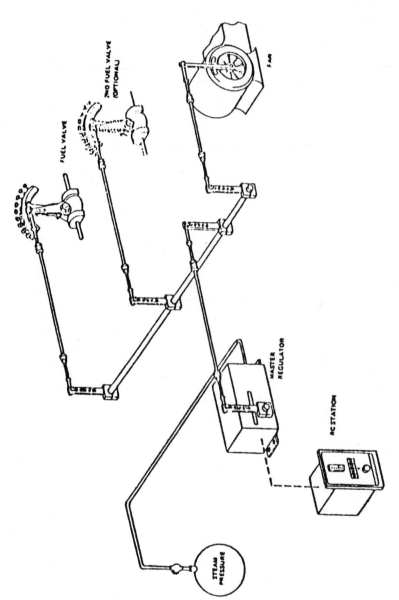

FIGURE 1-13. Typical parallel positioning type combustion control system using mechanical jackshaft.

Improved control of excess air can be obtained by substituting electric or pneumatic systems for the mechanical linkages. In addition, relative position of fuel control and combustion air dampers can be modified.

More advanced systems are pressure ratio control of the fuel and air pressure, direct air and fuel metering and excess air correction systems using flue gas O_2 monitoring. Factors that have limited the application of the most sophisticated control systems to industrial boilers include cost, reliability and maintenance.

2

Boiler Efficiency Goals

There are several ways of defining "boiler efficiency":

As-Found Efficiency is the efficiency measured in the field for boilers existing in a state of repair or maintenance. It is used as the baseline for any subsequent efficiency improvements.

Tuned-Up Efficiency is the efficiency after operating adjustments (low excess air) and minor repairs have been made.

Maximum Attainable Efficiency is the result of adding currently available efficiency improvement equipment, regardless of the cost considerations.

Maximum Economically Achievable Efficiency differs from that above in that it accounts for realistic cost considerations with efficiency improvement equipment added only if it is economically justifiable.

AS-FOUND EFFICIENCIES

As illustrated in Figure 2-1, there is a significant range of operating efficiencies dependent on the fuel fired and the existence of stack gas heat recovery equipment. The average efficiency ranges from 76% to 83% on gas, 78% to 89% on oil and 85% to 88% on coal. Note that the operating efficiency also varies with load.

Table 2-1 presents average "as-found" industrial boiler operating efficiency based on both field test measurements conducted by KVB and calculated values based on DOE data of industrial boilers. Good agreement is shown between the measured and calculated values.

21

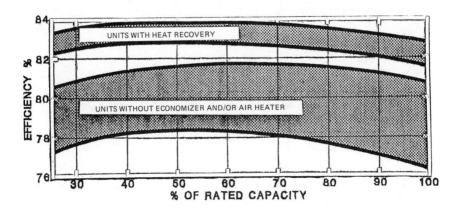

Typical Performance of Gas-Fired Watertube Boiler.

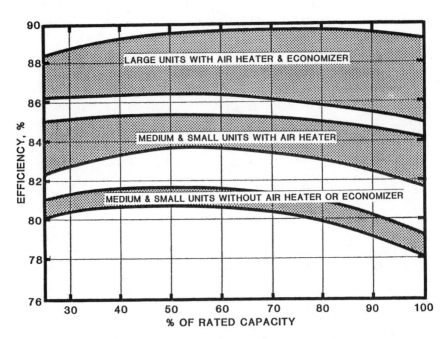

Typical Performance of Oil-Fired Watertube Boiler.

FIGURE 2-1. Ranges of Boiler Operating Efficiencies.

(continued)

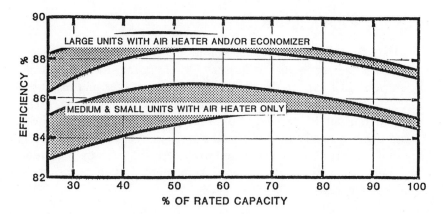

Typical Performance of Pulverized Coal-Fired Watertube Boiler.

(end)

TABLE 2-1. Average "As-Found" Industrial Boiler
Operating Efficiencies Field Test
Measurements and Calculations

(Percent)

| | Rated Capacity Range (MBtu/hr) | | | | | | | |
| | 10-16 | | 16-100 | | 100-250 | | 250-500 | |
Category/Fuel	Meas-ured	Calcu-lated	Meas-ured	Calcu-lated	Meas-ured	Calcu-lated	Meas-ured	Calcu-lated
Watertube								
Gas	(78.0)	79.9	79.5	79.9	81.2	80.9	(82.8)	81.2
Oil	(81.5)	83.7	82.8	83.7	(83.4)	84.6	(82.7)	85.3
Coal -Stoker	*	81.0	76.6	81.2	82.2	81.8	*	82.5
Pulverized	*	83.2	*	83.3	(86.6)	86.1	(85.3)	86.3
Firetube								
Gas	(81.0)	79.9	79.5	79.9	NA		NA	
Oil	(86.3)	83.7	(85.8)	83.7	NA		NA	

*No data available
NA - Not Applicable
Parentheses indicate small boiler populations tested

TUNED-UP EFFICIENCIES

Data from several programs conducted by KVB have been used to determine tuned-up efficiency levels using low excess air operation. In addition, boiler design efficiency is sometimes used as a reference point for establishing tuned-up efficiency.

Calculations of tuned-up boiler efficiency levels have also been made using typical tuned-up excess air levels as per manufacturers and data from the DOE. The results of these analyses are presented in Table 2-2. Again, reasonable agreement between the three methods are evident.

MAXIMUM
ECONOMICALLY ACHIEVABLE EFFICIENCIES

Several factors (as discussed in Chapter 18) are involved in the determination of the cost effectiveness of auxiliary equipment addition. Any estimate of the economic benefits that determine the economic feasibility of installing efficiency improvement equipment must be highly qualified due to the individual economic situation of each unit.

An analysis was conducted by KVB that showed that the addition of stack gas heat recovery equipment is the most cost-effective means of improving boiler efficiency. Table 2-3 presents the calculated maximum economically achievable efficiency levels based on the addition of stack gas heat recovery on units with sufficient potential to justify their addition.

MAXIMUM ATTAINABLE EFFICIENCY

The maximum attainable efficiency was calculated by KVB for each boiler category on the basis of applying all required auxiliary equipment to achieve minimum practical operating excess air levels and stack gas temperatures (see Chapter 3). These results are presented in Table 2-4 which show the expected trend of larger units having the highest efficiencies for each fuel and firing group due to lower radiation losses. Also, pulverized coal

TABLE 2-2. Average Measured, Design and Calculated Tuned-Up Industrial Boiler Operating Efficiencies

(Percent)

Category/Fuel	Rated Capacity Range, (10⁶ Btu/hr)											
	10-16			16-100			100-250			250-500		
	Meas-ured	Design	Calcu-lated	Meas-ured	Design	Calcu-lated	Meas-ured	Design	Calcu-lated	Meas-ured	Design	Calcu-lated
Watertube												
Gas	(78.4)	*	80.1	81.2	80.2	80.2	81.2	82.0	81.7	(83.7)	(83.4)	82.0
Oil	*	*	84.1	83.7	82.5	84.2	82.8	85.1	85.5	(81.5)	*	86.2
Coal - Stoker	*	*	81.3	80.9	(86.6)	81.0	83.6	(82.9)	82.0	*	*	82.7
Pulverized	*	*	83.8	*	*	83.9	(86.1)	(86.3)	86.5	(85.4)	(88.0)	86.7
Firetube												
Gas	(81.2)	(82.0)	80.1	(81.9)	(83.0)	80.2	NA	NA	NA	NA	NA	NA
Oil	(86.2)	(85.0)	84.1	(87.4)	(85.0)	84.2	NA	NA	NA	NA	NA	NA

* Insufficient data available
NA - Not Applicable
Numbers in parentheses represent small boiler population group.

TABLE 2-3. Calculated Maximum Economically Achievable
Efficiency Levels

(Percent)

Category/Fuel	Rated Capacity Range (10^6 Btu/hr)			
	10-16	16-100	100-250	250-500
Gas	80.1	81.7	84.0	85.2
Oil	84.1	86.7	88.3	88.7
Coal				
Stoker	81.0	83.9	85.5	85.8
Pulverized	83.3	86.8	88.8	89.1

TABLE 2-4. Calculated Maximum Attainable Efficiency Levels

(Percent)

	10-16 Klb/hr	16-100 Klb/hr	100-250 Klb/hr	250-500 Klb/hr
Gas	85.6	86.2	86.5	86.6
Oil	88.8	89.4	89.7	89.8
Coal				
Stoker	86.4	87.0	87.3	87.4
Pulverized	89.5	90.1	90.4	90.5

has the highest efficiency with oil and gas following. This ranking
follows the fuel properties as discussed in Chapter 3.

A summary of the efficiency levels and potential improvements from normal operating conditions is presented in Table 2-5.

$$\text{Fuel savings } = \frac{\text{Efficiency change}}{\text{Improved efficiency}} \times 100$$

There exists a 0.2 to 0.9% efficiency improvement potential
between "as-found" and tuned-up conditions. This corresponds
very favorably with the demonstrated efficiency improvements
from field test programs conducted by KVB. In addition 1.5 to
3.0% improvement potential is available using economically justified auxiliary equipment.

TABLE 2-5. Industrial Boiler Energy Conservation Potential

10-16 k lb/hr

Fuel	Baseline η	Tuned Up Δη	Tuned Up %Fuel	Max. Econ. Attainable η	Max. Econ. Attainable Δη	Max. Econ. Attainable %Fuel	Maximum Attainable η	Maximum Attainable Δη	Maximum Attainable %Fuel
Gas	79.9	0.2	0.25	80.1	0.2	0.25	85.6	5.7	6.66
Oil	83.7	0.4	0.48	84.1	0.4	0.48	88.8	5.1	5.74
Coal - Stoker	81.0	0.3	0.37	81.3	0.3	0.37	86.4	5.3	6.13
Pulverized	83.2	0.6	0.72	83.8	0.6	0.72	89.5	6.3	7.04

100-250 k lb/hr

Fuel	Baseline η	Tuned Up Δη	Tuned Up %Fuel	Max. Econ. Attainable η	Max. Econ. Attainable Δη	Max. Econ. Attainable %Fuel	Maximum Attainable η	Maximum Attainable Δη	Maximum Attainable %Fuel
Gas	80.9	0.8	0.98	84.0	3.1	3.69	86.5	5.6	6.47
Oil	84.6	0.9	1.05	88.3	3.7	4.19	89.7	5.1	5.69
Coal - Stoker	81.8	0.2	0.24	85.5	3.7	4.33	87.3	5.5	6.30
Pulverized	86.1	0.4	0.46	88.8	2.7	3.04	90.4	4.3	4.76

10-100 k lb/hr

Fuel	Baseline η	Tuned Up η	Tuned Up Δη	Tuned Up %Fuel	Max. Econ. Attainable η	Max. Econ. Attainable Δη	Max. Econ. Attainable %Fuel	Maximum Attainable η	Maximum Attainable Δη	Maximum Attainable %Fuel
Gas	79.9	80.2	0.3	0.37	81.7	1.8	2.20	86.2	6.3	7.31
Oil	83.7	84.2	0.5	0.59	86.7	3.0	3.46	89.4	5.7	6.38
Coal - Stoker	81.2	81.9	0.7	0.86	83.9	2.7	3.22	87.0	5.8	6.67
Pulverized	83.3	83.9	0.6	0.72	86.8	2.5	2.88	90.1	6.8	7.55

250-500 k lb/hr

Fuel	Baseline η	Tuned Up η	Tuned Up Δη	Tuned Up %Fuel	Max. Econ. Attainable η	Max. Econ. Attainable Δη	Max. Econ. Attainable %Fuel	Maximum Attainable η	Maximum Attainable Δη	Maximum Attainable %Fuel
Gas	81.2	82.0	0.8	0.98	85.2	4.0	4.70	86.6	5.4	6.24
Oil	85.3	86.2	0.9	1.06	88.7	3.4	3.83	89.8	4.5	5.01
Coal - Stoker	82.5	82.7	0.2	0.24	85.8	3.3	3.85	87.4	4.9	5.61
Pulverized	86.3	86.7	0.4	0.46	89.1	2.8	3.14	90.5	4.2	4.64

The maximum attainable efficiency improvements range from 2.0-4.0%. Note that this potential generally increases for all fuel categories with decreasing unit capacity with the exception of the smallest size category. This indicates the absence of stack gas heat recovery equipment.

3

Major Factors Controlling Boiler Efficiency

Boiler efficiency can be summarized as the measure of the efficiency with which the heat input to the boiler (principally the higher heating value of the fuel) is converted to useful output (in the form of process steam). Improvements in steam generator efficiency result primarily from reductions in waste heat energy losses in the stack gases and expelled waste water. Procedures that reduce the mass flow and energy content of these flow streams directly benefit unit performance. Other losses occur from surface heat transfer to the atmosphere and incomplete combustion of the fuel.

The proper calculation of boiler efficiency requires a definition of the boiler "envelope" which isolates the components to be considered part of the boiler from those that are excluded. Figure 4-1 from the next chapter, taken from the ASME Power Test Code, shows equipment included within the envelope boundary designating the steam generating unit. Heat inputs and outputs crossing the envelope boundary are involved in the efficiency calculations. Apparatus is generally considered outside the envelope boundary when it requires an outside source of heat or where the heat exchanged is not returned to the steam generating unit.

The direct approach to improving boiler efficiency is to identify the losses, their relative magnitude, and then concentrate first on the dominant losses that are controlling degraded efficiency. Some of the more important losses are listed below followed by a discussion of their origin.

1. *Waste heat energy losses in the stack gases* consist of the dry flue gas loss (heat carried away by the dry flue gases) and the moisture loss (latent and sensible heat in water vapor). Water vapor results from the combustion of hydrogen in the fuel, the humidity of the combustion air, and the water content of the fuel.

Most industrial boilers have very large flue gas losses because they operate with high stack gas temperatures (400°F-600°F+) resulting from not being equipped with waste heat recovery equipment (air preheater or economizers).

Traditionally, these boilers have not had the sophisticated combustion controls common on large utility boilers and as a result also operate with high dry flue gas losses due to the high excess air levels (20%-60%) necessary to insure complete combustion and safe operation. The latent heat of water vapor usually comprises a large fraction (6-10%) of the total efficiency losses and could be reduced if a practical means were developed to permit the water vapor to condense out before the flue gases leave the boiler.

In examining these stack gas efficiency losses, it is apparent that any reduction in the exit flue gas temperature and excess air level will help optimize the overall unit efficiency. A 100°F reduction in stack gas temperature will increase efficiency by 2.1% or more depending on the actual excess air levels (see Figure 8-2 in Chapter 8).

Minimum flue gas temperatures are limited by corrosion and sulfuric acid condensation in the cold end regions of the unit and are therefore a function of the sulfur content of the fuel and the moisture of the flue gas. One manufacturer of heat recovery equipment suggests a minimum average cold end temperature of 150°F for natural gas, 175°F for oil fuel and 155-185°F for coal, depending on the sulfur level of the fuel.

For boilers without heat recovery equipment, the minimum exit gas temperature is fixed by the boiler operating pressure since this determines the steam temperature. Usual design practices result in an outlet gas temperature ∿150°F above saturation temperature. Figure 3-1 illustrates the fact that it becomes increasingly expensive to approach boiler saturation temperatures by simply adding convective surface area. As operating pressures

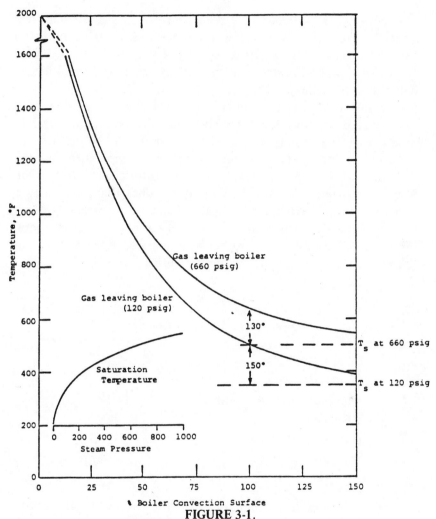

FIGURE 3-1.
Gas temperature drop through boiler convection section.

increase, the stack gas temperature increases making heat recovery equipment more desirable.

The practical limit for the minimum excess air is determined by the combustion control system used to regulate the air and fuel supply in response to load demand.

Economizers will permit a reduction in exit gas temperatures since the feedwater is at a lower temperature (220°F) than the steam saturation temperature. Stack gas temperatures of 300°F can be achieved with stack gas heat recovery equipment. Further reductions are achieved using air preheaters.

Present design criteria limits the degree of cooling using stack gas heat recovery equipment to a level which will minimize condensation on heat transfer surfaces. The sulfur content of the fuel has a direct bearing on the minimum stack gas temperature as SO_3 combines with condensed water to form sulfuric acid and also the SO_3 concentration in the flue gas determines the condensation temperature. Minimum air preheater metal surface temperatures are determined by averaging the exit gas and entering air temperature as given in Table 3-1. As shown, increased sulfur content in the fuel requires higher exit gas temperatures.

**TABLE 3-1. Minimum Air Preheater Exit Gas
Temperatures for 80°F Entering Air.**

Oil Fuel (>2.5% S)	390°F
Oil Fuel (<1.0% S)	330°F
Bituminous Coal (>3.5% S)	290°F
Bituminous Coal (<1.5% S)	230°F
Pulverized Anthracite	220°F
Natural Gas (Sulfur-free)	220°F

Table 3-2 presents a summary of the fuel properties and resulting minimum stack gas temperature and excess air levels for various fuel/firing types and the resulting total stack gas losses. Increases in stack gas temperature and excess air levels due to the factor discussed previously will result in increased losses.

2. *Another source of efficiency loss in a boiler that is not operating properly or is poorly maintained is the loss due to incomplete combustion.* This includes the loss due to combustible material in the flue gases (carbon monoxide, hydrogen, carbon

TABLE 3-2. Influence of Fuel Properties on Stack Gas Losses

Fuel	Firing Type	Fuel H (% wt)	Fuel H_2O (% wt)	Fuel S (% wt)	Min. Stack Temp (°F)	Min. Excess O_2 Level (%)	Min. Dry Gas Losses	Min. Moist Losses	Min. Stack Gas Losses
Natural gas		23.3	—	—	220	1	2.9	10.1	13.0%
#2 Oil		11.9	—	0.5	330	2	5.1	6.4	11.5%
#6 Oil		10.8	—	1.5	390	3	6.6	6.2	12.6%
Bituminous coal	Pulverized	5.0	4.5	2.8	290	4	4.8	4.5	9.3%
	Stoker	5.0	4.5	2.8	290	6	5.5	4.5	10.0%
Subbituminous coal	Pulverized	4.1	21.5	0.8	230	4	2.6	6.8	9.5%
	Stoker	4.1	21.5	0.8	230	6	3.0	6.8	9.8%

carryover, hydrocarbons, and smoke), and the loss due to un-burned solid fuel and other combustible solids which become trapped in the refuse.

As previously mentioned, the high excess air levels frequently used tend to minimize this loss unless the boiler is improperly maintained, is an older, poorly designed, coal unit, or is burning an uncommon fuel of inconsistent quality. High carbon monoxide (CO) emissions may be encountered on gas fuel because the boiler is operated at too low an excess air level and these poor combustion conditions are not visually apparent to the operator.

While combustible losses at gas- and oil-fired boilers can be essentially eliminated with proper operating practices, combustible losses on coal burning units are to some extent unavoidable. For coal the magnitude of the loss is very dependent upon the firing type (i.e., pulverized, stoker, cyclone).

These losses are evident by the combustible content of the ash. The combustible loss on pulverized coal units is dependent upon a number of variables including: (1) furnace heat liberation, (2) type of furnace cooling, (3) slag tap or dry ash removal, (4) volatility and fineness of coal, (5) excess air, (6) burner type, (7) burner-to-burner combustion balance and others. The typical combustible loss for each firing type is presented in Table 3-3.

TABLE 3-3. Typical Combustible Loss by Firing Type

Firing Type	Typical Combustible Losses (%)
Pulverized Coal	
Slag tap furnace	0
Cyclone combustor	0
Dry ash furnace	.8-1.2
Stokers	
Underfed	2.0
Overfed	2.0
Spreader (70% ash recovery)	
Dumping grate	3.6
Reciprocating grate	3.0
Vibrating grate	3.0
Traveling grate	2.4

Figure 3-2 illustrates how the various efficiency losses are affected by changes in boiler excess O_2. While these measurements were made on natural gas, the general trends are also representative of oil- and coal-fired boilers (actual values will differ with fuels, burner designs, operating conditions, etc.)

Dry flue gases increase linearly with increased excess O_2 due to both higher massflow rates and higher stack temperatures. Combustibles (carbon monoxide) increase dramatically as the excess O_2 is decreased below an acceptable minimum point.

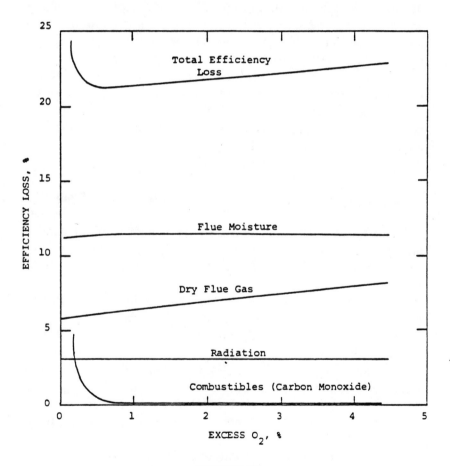

FIGURE 3-2.
Variation in boiler efficiency losses with changes in excess O_2.

Flue moisture and radiation losses remain unchanged with variation in excess O_2.

The total efficiency loss (the sum of the four sources listed above) decreases with decreased excess O_2 to a point in which losses from the combustibles become predominant. The optimum operating condition is not necessarily the point of highest efficiency due to additional excess O_2 margin required for safety control limitations or load changes.

The optimum excess air level for the best boiler efficiency occurs when the sum of the loss due to incomplete combustion and the loss due to heat in the flue gases is a minimum. For the ideal case of rapid thorough mixing, the optimum air-fuel ratio is the stoichiometric air-fuel ratio.

However, excess air is required in all practical cases to increase the completeness of combustion, allow for normal variations in the precision of combustion controls, and insure satisfactory stack conditions with some fuels (i.e., non-visible plume to comply with air pollution regulations). The optimum excess air level will vary with fuel, burner, and furnace design.

3. *Heat loss from the exterior boiler surfaces through the insulation is generally termed "radiation loss"* and includes heat radiated to the boiler room and the heat picked up by the ambient air in contact with the boiler surfaces. Approximate radiation losses from furnace walls as developed by the ABMA are presented in Figures 3-3 and 3-4.

The quantity of heat lost in this manner in terms of Btu per hour is fairly constant at different boiler firing rates and as a result, becomes an increasingly higher percentage of the total heat losses at the lower firing rates.

As seen in Figures 3-3 and 3-4, the radiation loss at high firing rates varies from a fraction of one percent up to two percent, depending on the capacity of the boiler. As the boiler load is reduced, the radiation loss increases in indirect proportion to the load fraction.

For example, the radiation loss for a 10,000 lb/hr boiler operating at 20% load will be five times the loss at full load, or roughly 10 percent. To a large extent, these losses are unavoidable

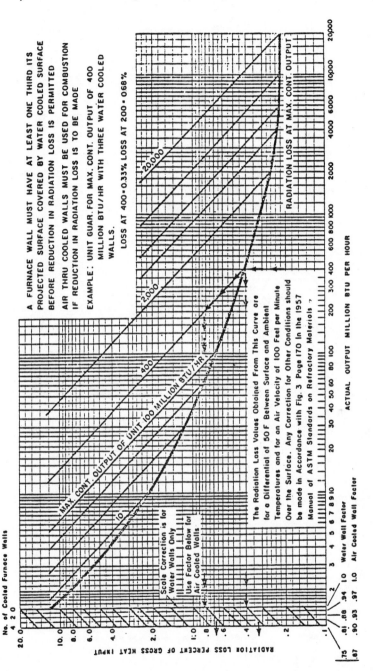

FIGURE 3-3. ABMA Standard Radiation Loss Chart.

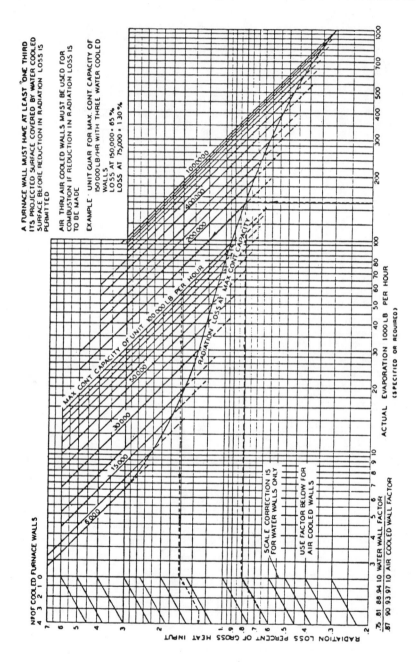

FIGURE 3-4. ABMA Standard Radiation Loss Chart.

and will increase at all loads with deteriorated insulation and furnace wall refractory. Techniques for reducing surface heat losses are presented in Chapter 14.

Boiler Firing Rate

Boiler firing rate is another operating parameter which affects efficiency but this parameter is often viewed as an uncontrollable factor depending on steam demand. As discussed in Chapter 12, load management can be an effective tool in some situations to minimize fuel use by maintaining boiler loading near peak efficiency conditions for as long as possible. The importance of load is illustrated in Figure 3-5 which shows how the various efficiency losses change with variations in boiler firing rate. These results on natural gas fuel are based on tests conducted on the same boiler where O_2 variations were previously discussed.

As indicated in the figure, the change in excess O_2 with load has a strong influence on the eventual efficiency versus load profile. When the boiler is fired with constant excess O_2 over the load range the actual peak efficiency may occur somewhat below peak load but the efficiency profile remains very "flat" over a large portion of the load range.

On the other hand, when excess O_2 increases as load is reduced (a common condition at many boilers), the efficiency tapers off more quickly with load. In this case it is advantageous to operate as close as possible to peak load for highest efficiencies when there is a choice between partially loading several boilers or operating fewer boilers at high loads.

A sample output from the KVB boiler efficiency computer program (Figure 3-6) shows the various heat losses and boiler efficiency at several test conditions on a 13,000 lb/hr watertube boiler. The radiation loss is based on the ABMA Standard Radiation Loss Chart mentioned previously. It should also be mentioned that no "unmeasured" or "unaccounted for" heat loss term has been applied to any efficiency values presented in this book. This additional loss term (generally ranging from 0.5 to 1.5 percent) is sometimes added to attempt to account for minor heat credits and losses which are neglected in the "short form" calculation.

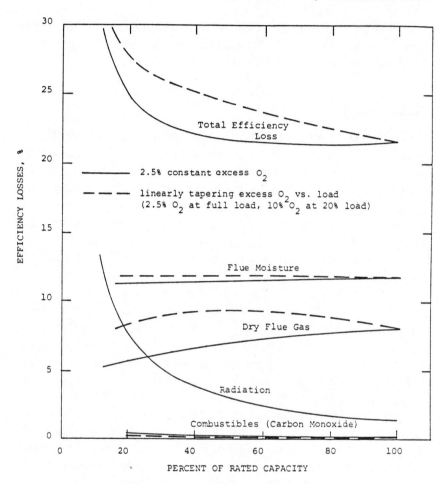

FIGURE 3-5. Variation in Boiler Efficiency Losses
With Changes in Boiler Firing Rate.

Effects of Boiler Operating Parameters

It will be worthwhile to examine in more detail the relative
importance of the various heat loss contributions and how they
vary with changes in boiler operating parameters. Figure 3-7
presents the results from a series of efficiency tests conducted at
a small watertube boiler while operating on natural gas fuel near
50 percent load capacity. To establish the optimum burner excess

O_2 condition for maximum efficiency, the combustion air flow was varied manually at a fixed fuel flow producing the range in excess O_2.

The lower plot shows the dependence of CO and flue gas temperature on excess O_2 while the upper plot shows the major heat losses at corresponding test points. The total efficiency loss profile exhibits a minimum value near 0.8% excess O_2 which corresponds to an efficiency peak of 78.8 percent. The total efficiency loss profile is shaped mainly by the dry flue gas losses and carbon monoxide losses since the radiation and moisture losses are nearly constant.

The point of maximum efficiency (minimum losses) occurs where the rate of change in CO and dry gas losses are equal and opposite. Since the dry gas loss continues to decrease very uniformly as the excess O_2 is lowered, the CO loss (i.e., the boilers' CO versus O_2 characteristics) is the primary factor in determining the point of maximum efficiency. In this particular example, the CO levels increase very rapidly below 1.0% excess O_2 and the point of maximum efficiency corresponds to CO emission levels in the region of 500 to 1000 ppm.

The importance of CO emissions in determining the point of peak efficiency is usually relevant to natural gas firing only. On oil and coal fuels the lowest excess O_2 is usually limited by an unacceptable stack condition (i.e., smoking) or excessive combustibles in refuse or fly ash. These conditions frequently precede high CO emissions but CO measurements are still made since CO can also result from malfunctioning burners, improper burner settings, etc.

It should be mentioned that the data in Figure 3-7 do not correspond to totally constant boiler output since fuel flow (input) is fixed and actual steam flow (output) would vary in proportion to the boiler efficiency. However, the efficiency loss profiles would be virtually unchanged if corrected to a constant steam flow condition. These curves illustrate the process of extracting heat from a given amount of fuel as opposed to the production of a constant quantity of steam.

FIGURE 3-6. Sample Computer Output from a Boiler Efficiency Program.

KVB Engineering   Calculation of Efficiency
Program: PEP3   Engineer: T. Sonnichsen
Boiler Category 111

Unit Description		Fuel Analysis		
		Natural Gas	Oil or Coal	
Location No.	ST JO	CO_2	.22	C	0.
Boiler No.	2	CO	0.	H	0.
Furnace Type	WT	N2	1.48	O	0.
Capacity		H2S	0.	N	0.
KLB/HR	13.0	CH4	92.88	S	0.
MBTU/HR	13.0	C_2H_6	4.17	H_2O	0.
Installed	UNK	C_3H_8	.93	ASH	0.
Erection Method	SHOP	C_4H_{10}	.19	HHV/(BTU/LB)	0
Burner Type	GUN	C_5H_{12}	.08		
		HHV(BTU/CUFT)	1055		

Natural Gas
Boiler Conditions

Test No.	5	6	7	8	X-6
Test Load					
KLB/HR	6.5	6.5	6.5	6.5	6.5
% of CAP	50.0	50.0	50.0	50.0	50.0

(continued)

Stack O$_2$ (% dry)	4.40	1.50	1.02	.55	.33
Stack CO (ppm)	70.0	360.0	580.0	1100.0	5000.0
Stack Temp (F)	476.0	460.0	455.0	458.0	457.0
AMB Air Temp (F)	80.0	80.0	80.0	80.0	80.0
Boiler Heat Balance Losses (%)					
Dry Gas	8.07	6.62	6.38	6.28	6.15
Moist + H$_2$	11.67	11.61	11.59	11.60	11.59
Moisture in Air	.21	.17	.17	.16	.16
Unburned CO	.02	.11	.17	.31	1.38
Combustibles	0.	0.	0.	0.	0.
Radiation	3.00	3.00	3.00	3.00	3.00
Boiler Efficiency	77.02	78.49	78.70	78.65	77.72
Thermodynamic Eff.	0.	0.	0.	0.	0.

(end)

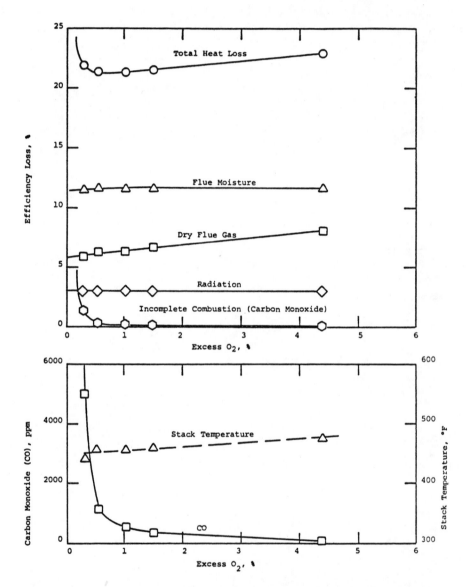

FIGURE 3-7. Variation in Stack Conditions and Heat Losses with
Changes in Excess O_2.

4

Boiler Efficiency Calculations

A necessary part of an efficiency improvement program is the determination of the operating efficiency of the boiler and the corresponding increase from "as-found" conditions. This chapter discusses the various calculation methods and computational procedures available. All are based on conducting an energy balance on the boiler system as illustrated in Figure 4-1.

A complete energy analysis would consider all such energy "credits" and "debits" (presented in Figure 4-2) to arrive at an overall efficiency. Generally, however, only the major factors are determined to arrive at an approximate value.

This chapter includes a discussion of the calculational methods available and the procedures developed by the ASME to conduct performance tests.

CALCULATION METHODS

The two basic procedures for determining the overall boiler efficiency are the input-output method and the heat loss method. Both methods are mathematically equivalent and would give identical results if all the required heat balance (or energy loss) factors were considered and the corresponding boiler measurements could be performed without error.

Different boiler measurements are required for each method. The efficiencies determined by these methods are "gross" efficiencies as opposed to "net" values which would include as additional heat input the energy required to operate all the boiler auxiliary equipment (combustion air fans, fuel handling systems, stoker

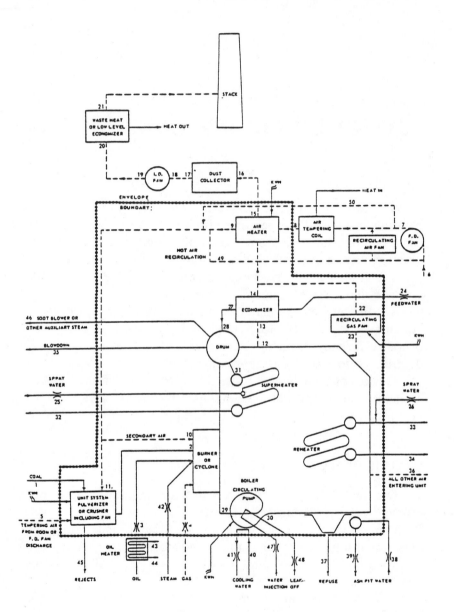

FIGURE 4-1. Steam generating unit diagram.

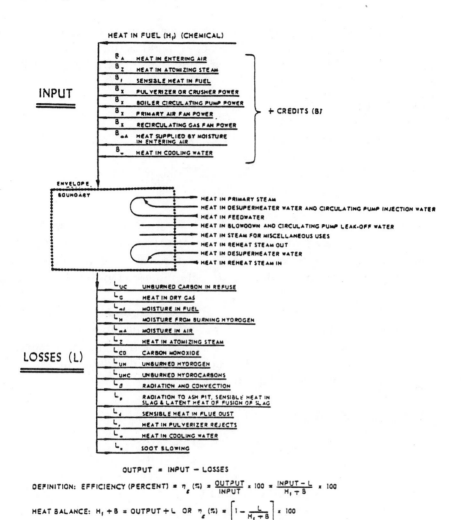

FIGURE 4-2. Heat balance of steam generator.

drives, etc.). These "gross" efficiencies can then be considered as essentially the effectiveness of the boiler in extracting the available heat energy of the fuel (i.e., transferring it to the working fluid).

Input-Output Method

$$\text{Efficiency (percent)} = \frac{\text{Output}}{\text{Input}} \times 100$$

$$= \frac{\text{Heat Absorbed by the Working Fluid or Fluids}}{\text{Heat in Fuel}} \times 100$$

This method requires the direct measurement of the fuel flow rate to determine the input rate of energy. The temperature, pressure and flow rate of the boiler feed water and generated steam must also be measured to determine the output rate of energy.

Because of the large number of physical measurements required at the boiler and the potential for significant measurement errors, the Input-Output Method is not practical for field measurements at the majority of industrial boiler installations where precision instrumentation is not available.

Heat Loss Method

$$\text{Efficiency (percent)} = 100 - \frac{\text{Heat Losses}}{\text{Heat Input}} \times 100$$

This method might also be termed the flue gas analysis approach since the major heat losses are based on the measured flue gas conditions at the boiler exit together with an analysis of the fuel composition as discussed in Chapter 3. Radiation loss, which is not associated with flue gas conditions, is routinely estimated from standard curves as given in Figures 3-3 and 3-4, Chapter 3.

This method requires the determination of the stack gas excess O_2 (or CO_2), CO, combustibles, temperature and the combustion air temperature.

The heat loss method is a much more accurate and more accepted method of determining boiler efficiencies in the field provided that the measurements of the stack gas conditions described above are accurate and not subject to air dilution or flue gas flow stratification (as discussed in Chapter 6).

ASME COMPUTATIONAL PROCEDURES

Boiler efficiency calculations using the calculation methods discussed previously are based on the ASME abbreviated efficiency tests or so-called "short form" computation from the ASME Power Test Code 4.1, Figure 4-3. This is a recognized standard approach for routine efficiency testing in the field, especially at industrial boiler installations where instrumentation quite often is very minimal.

This computational procedure neglects minor efficiency losses and heat credits and considers only the chemical heat (higher heating value) in the fuel as energy input. The same test form is used for both the input-output and the heat loss methods. The data required for each of these methods as per the ASME is identified on the following test forms. Note that as complete data as possible should be taken to fully document the test results no matter which procedure is used.

Input-Output Method

The input-output efficiency is determined as Item 64. The total energy output (Item 31) equals the sum of (A) of the actual water evaporation (Item 26) times the net enthalpy increase (Item 16 - Item 17) and the heat output in blowdown water. Note that reheat is not used on industrial boilers. The total heat input (Item 29) equals the rate of fuel firing (Item 28) times the as fired higher heating value of the fuel (Item 41).

ASME Heat Loss Method

The heat loss efficiency is determined as Item 72 which equals 100 less the sum of the major heat losses as percent of total heat input listed as Items 65 to 70. Each of these heat losses is calculated by first determining the heat loss per pound of fuel, followed by conversion to a percent loss by the fuel heating value.

ASME TEST FORM

SUMMARY SHEET   FOR ABBREVIATED EFFICIENCY TEST   PTC 4.1-a (1964)

				TEST NO.		BOILER NO.	DATE	
OWNER OF PLANT				LOCATION				
TEST CONDUCTED BY				OBJECTIVE OF TEST			DURATION	
BOILER MAKE & TYPE				RATED CAPACITY				
STOKER TYPE & SIZE								
PULVERIZER, TYPE & SIZE				BURNER, TYPE & SIZE				
FUEL USED	MINE			COUNTY	STATE		SIZE AS FIRED	

	PRESSURES & TEMPERATURES					FUEL DATA			
1	STEAM PRESSURE IN BOILER DRUM	psia			COAL AS FIRED PROX. ANALYSIS	%		OIL	
2	STEAM PRESSURE AT S. H. OUTLET	psia	I/O	37	MOISTURE	HL	51	FLASH POINT F°	
3	STEAM PRESSURE AT R. H. INLET	psia	I/O	38	VOL MATTER		52	Sp. Gravity Deg. API°	
4	STEAM PRESSURE AT R. H. OUTLET	psia	I/O	39	FIXED CARBON		53	VISCOSITY AT SSU° BURNER SSF	
5	STEAM TEMPERATURE AT S. H. OUTLET	F	I/O	40	ASH		44	TOTAL HYDROGEN % wt	
6	STEAM TEMPERATURE AT R H INLET	F	I/O		TOTAL		41	Btu per lb	
7	STEAM TEMPERATURE AT R. H. OUTLET	F	I/O	41	Btu per lb AS FIRED	I/O and HL			
8	WATER TEMP. ENTERING (ECON.)(BOILER)	F	I/O	42	ASH SOFT TEMP.° ASTM METHOD			GAS	% VOL
9	STEAM QUALITY % MOISTURE OR P. P. M.		I/O		COAL OR OIL AS FIRED ULTIMATE ANALYSIS		54	CO	
10	AIR TEMP. AROUND BOILER (AMBIENT)	F		43	CARBON	HL	55	CH₄ METHANE	
11	TEMP AIR FOR COMBUSTION (This is Reference Temperature) †	F	HL	44	HYDROGEN	HL	56	C₂H₂ ACETYLENE	
12	TEMPERATURE OF FUEL	F		45	OXYGEN		57	C₂H₄ ETHYLENE	
13	GAS TEMP. LEAVING (Boiler) (Econ.) (Air Htr.)	F	HL	46	NITROGEN		58	C₂H₆ ETHANE	
14	GAS TEMP. ENTERING AH (If conditions to be corrected to guarantee)	F		47	SULPHUR	HL	59	H₂ S	
	UNIT QUANTITIES			40	ASH		60	CO₂	
15	ENTHALPY OF SAT. LIQUID (TOTAL HEAT)	Btu/lb	I/O	37	MOISTURE		61	H₂ HYDROGEN	
16	ENTHALPY OF (SATURATED) (SUPERHEATED) STM.	Btu/lb	I/O		TOTAL			TOTAL	
17	ENTHALPY OF SAT. FEED TO (BOILER) (ECON.)	Btu/lb	I/O		COAL PULVERIZATION			TOTAL HYDROGEN % wt	
18	ENTHALPY OF REHEATED STEAM R.H. INLET	Btu/lb	I/O	48	GRINDABILITY INDEX°		62	DENSITY 68 F ATM. PRESS.	
19	ENTHALPY OF REHEATED STEAM R. H. OUTLET	Btu/lb	I/O	49	FINENESS % THRU 50 M°		63	Btu PER CU FT	
20	HEAT ABS/LB OF STEAM (ITEM 16 – ITEM 17)	Btu/lb	I/O	50	FINENESS % THRU 200 M°		41	Btu PER LB	
21	HEAT ABS/LB R. H. STEAM (ITEM 19 – ITEM 18)	Btu/lb	I/O	64	INPUT-OUTPUT EFFICIENCY OF UNIT %		ITEM 31 = 100 ITEM 29	I/O	
22	DRY REFUSE (ASH PIT + FLY ASH) PER LB AS FIRED FUEL	lb/lb	HL					Btu/lb A. F. FUEL	% of A. F. FUEL
23	Btu PER LB IN REFUSE (WEIGHTED AVERAGE)	Btu/lb	HL		HEAT LOSS EFFICIENCY				HL
24	CARBON BURNED PER LB AS FIRED FUEL	lb/lb	HL	65	HEAT LOSS DUE TO DRY GAS				HL
25	DRY GAS PER LB AS FIRED FUEL BURNED	lb/lb	HL	66	HEAT LOSS DUE TO MOISTURE IN FUEL				HL
	HOURLY QUANTITIES			67	HEAT LOSS DUE TO H₂O FROM COMB. OF H₂				HL
26	ACTUAL WATER EVAPORATED	lb/hr	I/O	68	HEAT LOSS DUE TO COMBUST. IN REFUSE				HL
27	REHEAT STEAM FLOW	lb/hr	I/O	69	HEAT LOSS DUE TO RADIATION				
28	RATE OF FUEL FIRING (AS FIRED wt)	lb/hr	I/O	70	UNMEASURED LOSSES				
29	TOTAL HEAT INPUT (Item 28 x Item 41)/1000	kB/hr	I/O	71	TOTAL				HL
30	HEAT OUTPUT IN BLOW-DOWN WATER	kB/hr		72	EFFICIENCY = (100 – item 71)				HL
31	TOTAL HEAT OUTPUT (Item 26 x Item 20) + (Item 27 x Item 21) + Item 30 /1000	kB/hr	I/O						
	FLUE GAS ANAL. (BOILER) (ECON) (AIR HTR) OUTLET								
32	CO₂	% VOL	HL		I/O – Input - Output Method				
33	O₂	% VOL	HL		HL – Heat Loss Method				
34	CO	% VOL	HL						
35	N₂ (BY DIFFERENCE)	% VOL	HL		° Not Required for Efficiency Testing				
36	EXCESS AIR	%	HL		† For Point of Measurement See Par. 7.2.8.1-PTC 4.1-1964				

FIGURE 4-3. ASME Short Form.

ASME TEST FORM

CALCULATION SHEET FOR ABBREVIATED EFFICIENCY TEST PTC 4.1-b (1964)

	OWNER OF PLANT	TEST NO.	BOILER.NO.	DATE

30 HEAT OUTPUT IN BOILER BLOW-DOWN WATER = LB OF WATER BLOW-DOWN PER HR x $\dfrac{\text{ITEM 15} \quad \text{ITEM 17}}{1000}$ = ... kB/hr

24 If impractical to weigh refuse, this item can be estimated as follows

DRY REFUSE PER LB OF AS FIRED FUEL = $\dfrac{\% \text{ ASH IN AS FIRED COAL}}{100 - \% \text{ COMB. IN REFUSE SAMPLE}}$

CARBON BURNED PER LB AS FIRED FUEL = $\dfrac{\text{ITEM 43}}{100} - \left[\dfrac{\text{ITEM 22}}{14,500} \times \text{ITEM 23}\right]$ = ...

NOTE: IF FLUE DUST & ASH PIT REFUSE DIFFER MATERIALLY IN COMBUSTIBLE CONTENT, THEY SHOULD BE ESTIMATED SEPARATELY. SEE SECTION 7, COMPUTATIONS.

25 DRY GAS PER LB AS FIRED FUEL BURNED = $\dfrac{11CO_2 + 8O_2 + 7(N_2 + CO)}{3(CO_2 + CO)}$ x (LB CARBON BURNED PER LB AS FIRED FUEL + $\frac{3}{8}$ S)

$= \dfrac{11 \times \overline{\text{ITEM 32}} + 8 \times \overline{\text{ITEM 33}} + 7\left(\overline{\text{ITEM 35}} + \overline{\text{ITEM 34}}\right)}{3 \times \left(\overline{\text{ITEM 32}} + \overline{\text{ITEM 34}}\right)} \times \left[\overline{\text{ITEM 24}} + \overline{\text{ITEM 47}}\right]$ = ...

36 EXCESS AIR † = 100 x $\dfrac{O_2 - \frac{CO}{2}}{.2682N_2 - (O_2 - \frac{CO}{2})}$ = 100 x $\dfrac{\text{ITEM 33} - \frac{\text{ITEM 34}}{2}}{.2682(\text{ITEM 35}) - (\text{ITEM 33} - \frac{\text{ITEM 34}}{2})}$ = ...

HEAT LOSS EFFICIENCY	Btu/lb AS FIRED FUEL	$\frac{\text{LOSS}}{\text{HHV}} \times 100$ =	LOSS %
65 HEAT LOSS DUE TO DRY GAS = $\frac{\text{LB DRY GAS}}{\text{PER LB AS FIRED FUEL}} \times C_p \times (t_{vg} - t_{air})$ = $\text{ITEM 25} \times 0.24 \times \frac{(\text{ITEM 13}) - (\text{ITEM 11})}{\text{Unit}}$ = ...	......	$\frac{65}{41} \times 100$ =	......
66 HEAT LOSS DUE TO MOISTURE IN FUEL = $\frac{\text{LB H}_2\text{O PER LB}}{\text{AS FIRED FUEL}} \times$ [(ENTHALPY OF VAPOR AT 1 PSIA & T GAS LVG) − (ENTHALPY OF LIQUID AT T AIR)] = $\frac{\text{ITEM 37}}{100} \times$ [(ENTHALPY OF VAPOR AT 1 PSIA & T ITEM 13) − (ENTHALPY OF LIQUID AT T ITEM 11)] = ...	......	$\frac{66}{41} \times 100$ =	......
67 HEAT LOSS DUE TO H$_2$O FROM COMB. OF H$_2$ = 9H$_2$ × [(ENTHALPY OF VAPOR AT 1 PSIA & T GAS LVG) − (ENTHALPY OF LIQUID AT T AIR)] = 9 x $\frac{\text{ITEM 44}}{100}$ x [(ENTHALPY OF VAPOR AT 1 PSIA & T ITEM 13) − (ENTHALPY OF LIQUID AT T ITEM 11)] = ...	......	$\frac{67}{41} \times 100$ =	......
68 HEAT LOSS DUE TO COMBUSTIBLE IN REFUSE = $\frac{\text{ITEM 22}}{......} \times \frac{\text{ITEM 23}}{......}$ = ...	......	$\frac{68}{41} \times 100$ =	......
69 HEAT LOSS DUE TO RADIATION* = $\frac{\text{TOTAL BTU RADIATION LOSS PER HR}}{\text{LB AS FIRED FUEL}} - \text{ITEM 28}$ = ...	......	$\frac{69}{41} \times 100$ =	......
70 UNMEASURED LOSSES **	......	$\frac{70}{41} \times 100$ =	......
71 TOTAL	......	......	......
72 EFFICIENCY = (100 − ITEM 71)	......	......	......

† For rigorous determination of excess air see Appendix 9.2 − PTC 4.1-1964
* If losses are not measured, use ABMA Standard Radiation Loss Chart, Fig. 8, PTC 4.1-1964
** Unmeasured losses listed in PTC 4.1 but not tabulated above may by provided for by assigning a mutually agreed upon value for Item 70.

FIGURE 4-3. Continued.

The heat loss due to dry gas equals the pounds of dry gas per pound of fuel (Item 25) times the specific heat of the combustion gases (approximately 0.24 Btu/lb °F) times the temperature difference between the exit gas (Item 13) and the inlet air for combustion (Item 11). Note that the exit gas temperature will depend on the stack gas heat recovery equipment (economizer and/or air preheater) on the unit. The heat loss due to moisture in the fuel equals the pounds of water per pound of fuel times the enthalpy difference between the water vapor at exit gas temperature (Item 13) and water at ambient temperature (Item 11). The heat loss due to water vapor formed from hydrogen in the fuel equals the weight fraction of hydrogen in the fuel times 9 (there are approximately 9 pounds of water produced from burning one pound of hydrogen) times the enthalpy difference between the water vapor in the stack and the liquid at ambient temperature. The heat loss due to combustibles in the refuse equals the pounds of dry refuse per pound of fuel (Item 22) times the heating value of the refuse (Item 23) determined in a laboratory. The heat loss due to radiation is determined by the AMBA chart discussed previously. An unmeasured loss factor (generally ranging from 0.5 to 1.5 percent) is sometimes added to attempt to account for minor heat credits and losses which are neglected in the "short form" calculation.

Comparison of the
Input-Output and the Heat Loss Methods

Given in Table 4-1 is a comparison between the input-output and heat loss methods discussed in the previous sections. As shown, there is reasonably good agreement between the two calculation procedures.

However, for practical boiler tests with limited on-site instrumentation, comparisons between the two methods are generally poor, resulting primarily from the inaccuracies associated with the measurement of the flow and energy content of the input and output streams.

TABLE 4-1. Efficiency Calculation Comparison
#2 Oil

	ANL, Test #6
Nominal Boiler Conditions	
Steam Flow (lb/hr)	50,000
Final Steam Pressure (psig)	196
Boiler Excess Air	57%
Input/Output Efficiency	
Input Factors:	
Fuel Flow (lb/hr)	2,980
High Heating Value (Btu/lb)	19,400
Output Factors:	
Feedwater (Steam) Flow (lb/hr)	50,000
Enthalpy of Steam (Btu/lb)	1,198
Enthalpy of Feedwater	198
Heat Input (Btu/hr)	57,812
Heat Output (Btu/hr)	50,000
Input/Output Efficiency	86.5
Heat Loss Efficiency	
O_2 (%)	8.0
CO (ppm)	0
Temperature (°F)	275
Dry Gas Loss (%)	5.4
Moisture Loss (%)	6.2
Combustible Loss (%)	0
Radiation Loss (%)	0.9
Total Heat Loss (%)	12.9
Heat Loss Efficiency (%)	87.5

5

Heat Loss
Graphical Solutions

Relationships have been generated that can be used to estimate the percent of heat losses for conventional firing conditions that can be used in the heat loss computational procedures. These graphical solutions were developed by KVB using the ASME computational procedures discussed in Chapter 3. Solutions are presented in this chapter for the stack gas losses (dry flue gas and moisture losses combined) and the combustible losses.

Figures 5-1 through 5-5 can be used to estimate the stack gas losses for various operating conditions (flue gas excess O_2 and stack temperature) for natural gas, #2 and #6 oils, and an eastern bituminous and western subbituminous coals. The fuel analysis used as the basis for these solutions is presented and is characteristic of the general fuel properties.

Estimates of the total stack gas losses are made using the appropriate figure (based on fuel used) by determining both the flue gas excess O_2 and stack gas temperature using the measurement procedures presented in Chapter 6. The stack gas heat loss is read off the left side.

Figure 5-6 can be used to measure the heat loss from carbon monoxide present in the flue gas for natural gas firing. Required data are the CO emissions (in ppm) and the stack gas excess O_2 level.

To complete the heat loss estimate, Figure 4-3 is used to determine the heat loss due to radiation.

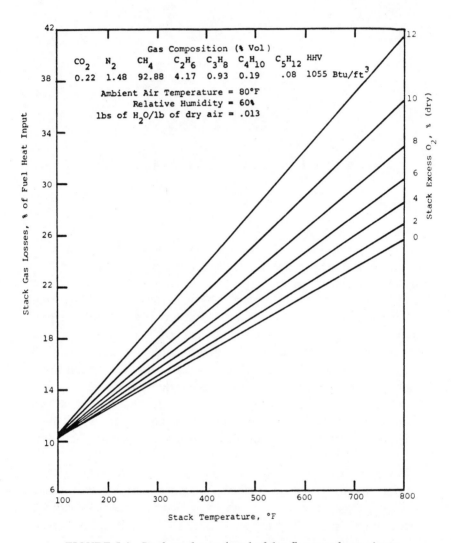

FIGURE 5-1. Stack gas losses (total of dry flue gas plus moisture
in air plus moisture in flue gas due to the combustion of
hydrogen in the fuel) as a function of stack temperature and
excess O_2 for natural gas fuel.

KVB®

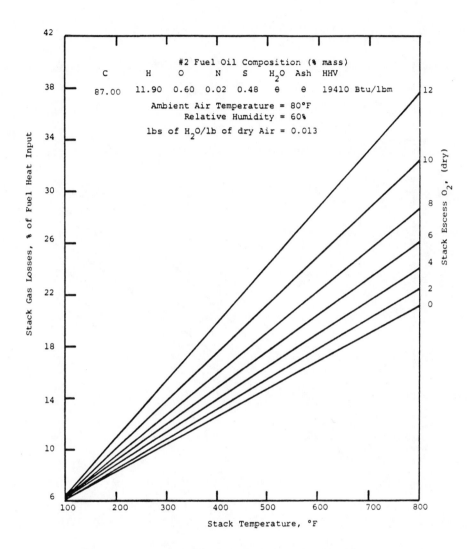

FIGURE 5-2. Stack gas losses (total of dry flue gas plus moisture in air plus moisture in flue gas due to the combustion of hydrogen in the fuel) as a function of temperature and excess O_2 for #2 fuel oil.

KVB®

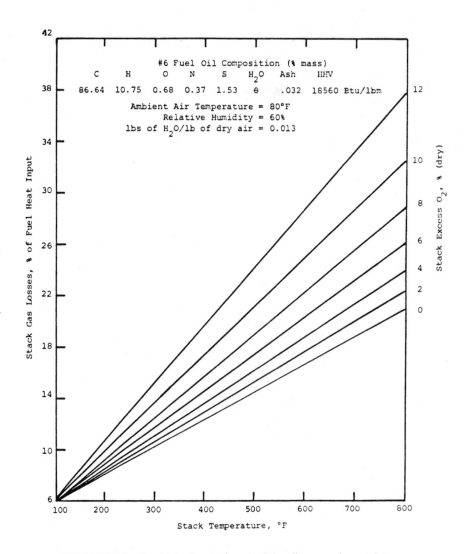

FIGURE 5-3. Stack gas losses (total of dry flue gas plus moisture in air plus moisture in flue gas due to the combustion of hydrogen in the fuel) as a function of stack temperature and excess O_2 for #5 and #6 fuel oils.

KVB®

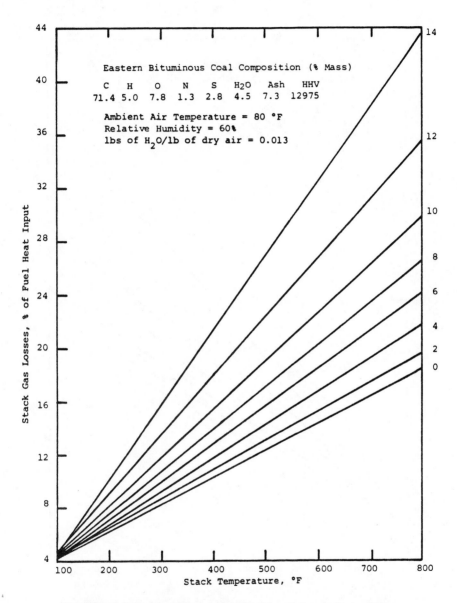

FIGURE 5-4. Stack gas losses (total of dry flue gas plus moisture in air plus moisture in flue gas due to the combustion of hydrogen in the fuel) as a function of stack temperature and excess O_2 for eastern bituminous coal.

KVB®

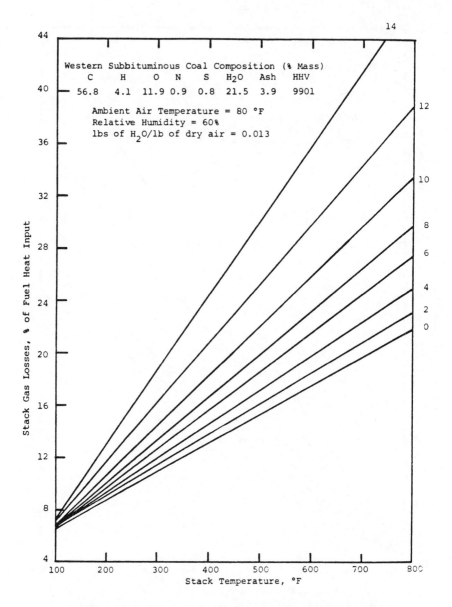

FIGURE 5-5. Stack gas losses (total of dry flue gas plus moisture
in air plus moisture in the flue gas due to the combustion of
hydrogen in the fuel) as a function of stack gas temperature
and excess O₂ for western subbituminous coal.

KVB®

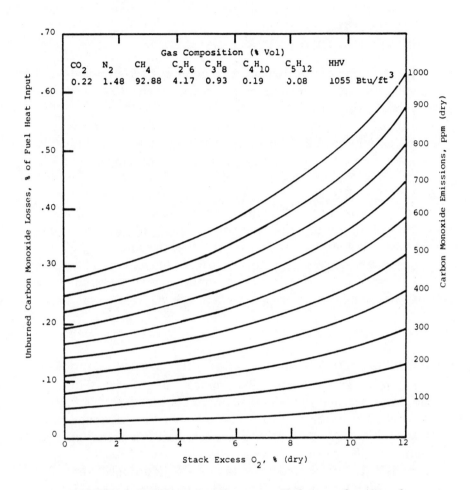

FIGURE 5-6. Unburned carbon monoxide loss as a function of excess O_2 and carbon monoxide emissions for natural gas fuel.

KVB®

ABBREVIATED EFFICIENCY
IMPROVEMENT DETERMINATIONS

The efficiency improvement for a boiler at each firing rate will depend on both the reductions in excess air and the stack temperatures.

Relationships giving the approximate efficiency improvements for reduction in these operating parameters are presented as Figures 5-8 and 5-9. These relationships were developed by KVB by assuming that the specific heat of the flue gases was nearly constant for all fuels and exit gas temperatures (consistent with ASME procedures) and that the stack gas heat losses are a linear function of excess air within the normal range of boiler operation.

These relationships can be used together to estimate efficiency improvements achieved at constant firing rates on natural gas, #2 through #6 oils, and common bituminous and subbituminous coals. An example of its use it presented below.

Reductions in Excess O_2

Figure 5-8 gives the approximate percent of efficiency improvement corresponding to each 1% reduction in excess air. Use Figure 1-1 to convert excess O_2 to excess air, and then determine the efficiency improvement at the proper stack temperature. For example, if the excess O_2 were reduced from 6% to 3.5% from Figure 1-1, it is determined that the excess air dropped from 40% to 20%, a 20% total change. If the stack temperature were 360°F, the efficiency improvement was 20 times 0.05 or 1.0 percent.

Reduction in Stack Gas Temperature

Reductions in stack temperature will also generally occur as the excess O_2 is lowered. To account for this effect, perform the previous calculation using the initial stack temperature and add to this improvement the efficiency improvement at the lower excess air level obtained from Figure 5-7.

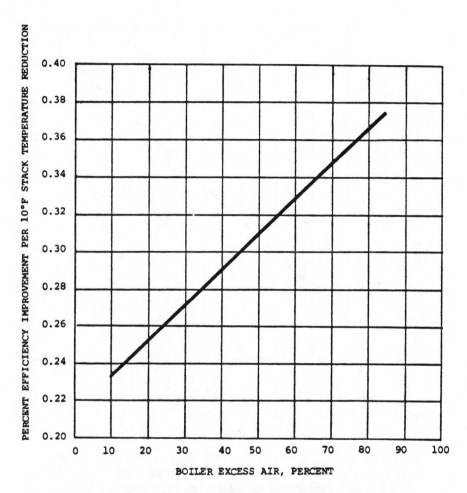

FIGURE 5-7. Curve showing percent efficiency improvement per every 10°F drop in stack temperature. Valid for estimating efficiency improvements on typical natural gas, #2 through #6 oils and coal fuels.

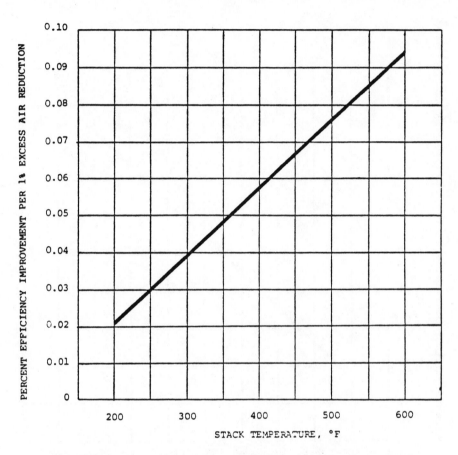

FIGURE 5-8. Curve showing percent efficiency improvement per every one percent reduction in excess air. Valid for estimating efficiency improvements on typical natural gas, #2 through #6 oils and coal fuels.

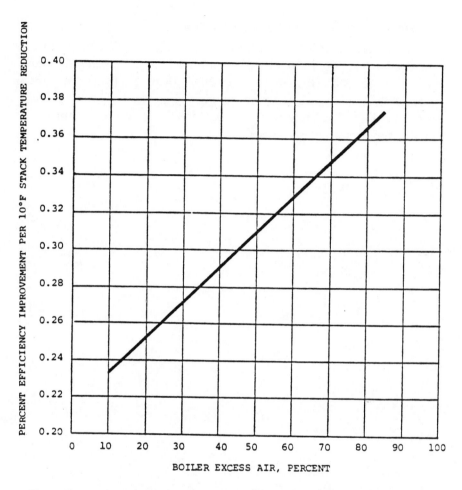

FIGURE 5-9. Curve showing percent efficiency improvement per every 10°F drop in stack temperature. Valid for estimating efficiency improvements on typical natural gas, #2 through #6 oils and coal fuels.

Example Use of The Abbrebiated Method

To illustrate, if in the previous example, the stack temperature dropped from 360°F to 340°F as the excess O_2 was lowered from 6% to 3.5%, the effect due to the 20°F lower stack temperature would be 2 times 0.25, which equals 0.5 percent. The total improvement in efficiency is then 1.0 plus 0.5, for a total improvement of approximately 1.5 percent.

6

Preparation for Boiler Testing

Efficiency improvements obtained under a deteriorated state of the boiler can be substantially less than the improvements achieved under proper working conditions. It is essential that the boiler be examined prior to testing and that necessary repairs or maintenance be completed. A partial summary of the items that should be included in the preliminary boiler inspection is given in Table 6-1. Essentially, any item that could ultimately affect the combustion process in the furnace should be included and repaired prior to implementing an efficiency improvement program.

STACK INSTRUMENTATION
Gaseous Constituents

It is necessary to measure the concentration of either excess O_2 or carbon dioxide to determine the operating excess air levels. (As discussed in Chapter 1, excess O_2 is generally preferred due to fewer measurement errors.)

Carbon monoxide is also measured. CO is the primary indicator of incomplete combustion on gas fuel. CO measurements on oil and coal fuel are generally not mandatory since smoking or excessive carbon carryover will usually precede high CO levels.

Portable electronic analyzers are available for both O_2 and CO_2 measurements. Alternative measurement techniques include an Orsat analyzer or other hand-held chemical absorbing type analyzer and "length of stain" detectors. Calibration procedures and verification are important criteria.

A partial list of the available portable analyzers is given in Table 6-2.

TABLE 6-1. Preliminary Boiler Inspection Checklist.

BURNERS				Combustion Controls	Furnace
Gas Firing	Oil Firing	Pulv. Coal Firing	Stoker Firing		
- Condition and cleanliness of gas injection orifices	- Condition and cleanliness of oil tip passages	- Condition and operation of pulverizers, feeders and conveyors	- Wear on grates	- Cleanliness and proper movement of fuel valves	- Excessive deposits or fouling of gas-side boiler tubes
- Cleanliness and operation of filter & moisture traps	- Oil burning temperature	- Coal fineness	- Position and operation of stokers	- Excessive "play" in control linkages or air dampers	- Proper operation of soot-blowers
- Condition and orientation of diffusers, spuds, gas canes, etc.	- Atomizing steam pressure	- Condition of coal pipes	- Positioning of all air proportioning dampers	- Adequate pressure to all pressure regulators	- Casing and duct leaks
- Condition of burner refractory	- Condition and orientation of burner diffusers	- For any signs of excessive erosion or burnoff	- Coal sizing	- Unnecessary cycling of firing rate	- Clean and operable furnace inspection ports
- Condition and operation of air dampers	- Position of oil guns	- Condition and operation of air dampers	- Operation of cinder reinjection system	- PROPER OPERATION OF ALL SAFETY INTERLOCKS AND BOILER TRIP CIRCUITS	
	- Cleanliness of oil strainer				
	- Condition of burner throat refractory				
	- Condition and operation of air dampers				

KVB®

TABLE 6-2. Portable Gaseous Stack Gas Analyzers Based on Manufacturer's Information (Incomplete).

Manufacturer	Model	Capabilities	Comments
Thermox Division, AMETEK, Inc.	WDG-P WDG-P-CA P-100	Oxygen Oxygen and Combustibles	Zirconium Oxide Measurement Zirconium Oxide and catalytic detector
Beckman	715	Oxygen	Auxiliary pump required
Teledyne	320P	Oxygen	Fuel cell detector Built-in pump
	980	Oxygen and Combustibles	Catalytic bed sensor, built-in pump, comb. calibration gas req.
Mine Safety Appliances (MSA)	830P	Oxygen	Electrolyte fuel cell detector, built-in pump
Bacharach	Combustion Testing Kit	Oxygen, Carbon Dioxide, Carbon Monoxide	Gaseous absorption for O_2 and CO_2, length-of-stain for CO; various additional test equipment available
Hays	621A.31:30	Oxygen, Carbon Dioxide, Carbon Monoxide	Orsat analyzer
Dragger	31	Carbon Monoxide	Length-of-stain analyzer

Stack Opacity (Smoke Density)

Smoking on oil or coal fuel is a certain indication of flue gas combustibles or unacceptable flame conditions and should always be avoided. Stack opacity is generally used as the criterion for determining minimum excess air levels for oil and coal firing (whereas excessive CO emissions are used on natural gas firing). Smoke measurements can be made using hand pump filter paper testers or visual observation to a Ringlemann scale.

Stack Temperature

Accurate stack gas temperature measurements can be obtained using a dial type temperature gage or thermocouples.

STACK SAMPLING TECHNIQUES

It is essential that the portion of the stack gas analyzed for temperature and gaseous constituents be a representative sample of the bulk of the stack gas flow. Examples of uniform stack gas conditions and severe maldistributions are given in Figures 6-1 and 6-2 respectively. Sample location should be selected so as to minimize the effects of air leakage and gaseous stratification in the exhaust duct.

FLAME APPEARANCE

The appearance of a boiler's flame can provide a good preliminary indication of combustion conditions. While the characteristics of a "good" flame are somewhat subjective, flames of a definite appearance have usually been sought. Oil and pulverized flames should be short, bright, crisp and highly turbulent. Gas flames should be blue, slightly streaked or nearly invisible.

For stokers, an even bed and an absence of carbon streams are important criteria. Stability of the flame at the burner and minimum furnace vibration are also desired.

Operation with reduced excess O_2 levels may result in a different flame appearance. Flames may appear to grow in volume

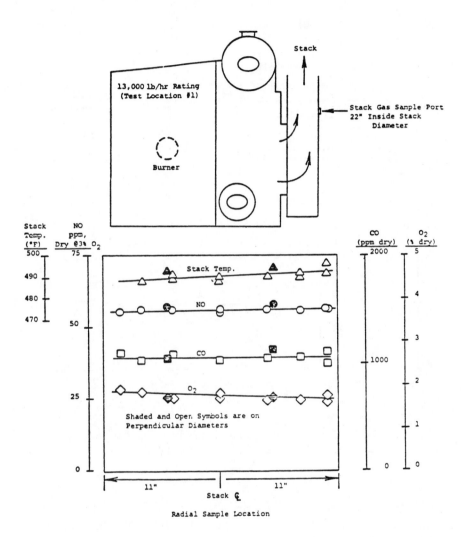

FIGURE 6-1. Flue gas composition and temperature profiles
at the outlet of a small D-type watertube boiler.

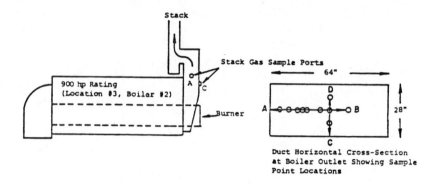

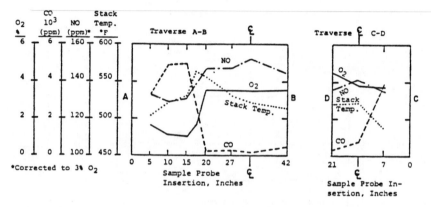

FIGURE 6-2. Flue gas composition and temperature profiles
at the outlet of a large firetube boiler.

and more completely "fill" the furnace. Low O_2 flames sometimes exhibit a "lazy" rolling appearance. The overall color may
change with natural gas fires becoming more visible or luminous
and coal and oil flames becoming darker yellow or orange.

7

An Update and Overview of Flue Gas Measurement

Timothy Jones, Product Manager
AMETEK, Inc., Thermox Instruments Division

Controlling the efficiency of combustion processes is as important today to the operators of large power boilers for an electrical utility as it is to the person who runs a small furnace or water heater in a residential dwelling. The steady increase of fossil fuel prices over the past two decades has made energy conservation an important way to control costs.

One of the best ways to reduce wasted fuel is to monitor—and increase—combustion efficiency. The amounts of oxygen and combustibles that are being allowed to flow out of a smokestack can be measured in a number of different ways. These measurements are facilitated by a new generation of microprocessor-based analyzers with self-diagnostic systems that are easier to calibrate and operate.

To make sure that the right gas analyzer is used for the job, plant operators and engineers must understand both the mechanics of combustion and why the amounts of excess oxygen and fuel leaving a stack should be kept to a minimum.

COMBUSTION EFFICIENCY

Combustion efficiency is a measurement of the effectiveness of a boiler, furnace or process heater to convert the energy within a fuel into heat. This efficiency can be expressed in an equation as the total energy contained per unit of fuel minus the energy carried away by the hot flue gases exiting up the stack divided by the total energy contained per unit of fuel:

$$\text{Percentage Efficiency} = \left(\frac{\left(\frac{Btu}{ft^3}\right) \text{fuel} - \left(\frac{Btu}{ft^3}\right) \text{flue gas}}{\left(\frac{Btu}{ft^3}\right) \text{fuel}} \right) \times 100$$

Refer back to Chapter 4 for a more detailed method of calculating efficiency.

Two key components of this efficiency equation are the stoichiometric air/fuel ratio and the heat of combustion value for the fuel being burned. The stoichiometric ratio shows the exact amounts of air and fuel needed for a combustible to be completely consumed. The heat of combustion shows the amount of energy that would be released in such a perfect reaction.

As an example, methane has a stoichiometric ratio of one cubic foot of methane to 9.53 cubic feet of air. (This is assuming the standard sample of air contains about 21 percent oxygen and 79 percent nitrogen.) At this ratio, the methane would be fully burned and would release 1013 Btu per cubic feet.

Therefore, any combustible that is burned at an air/fuel ratio which is higher or lower than its predetermined stoichiometric figure will result in either wasted fuel or wasted energy.

BURNING WITH EXCESS OXYGEN

Before energy efficiency was a concern, it was common to run a burner with large amounts of air to ensure that a fuel burned completely. Today, that is seen as a highly wasteful practice.

When there is too much air—or too little fuel—present during a burning process, only the amount of oxygen needed for stoichiometric combustion is used. The rest of the air simply flows up the flue, carrying with it useful heat from the process. This is because the air enters the burner at ambient temperature, but leaves at the increased temperature of the flue gas.

Energy loss from excess oxygen increases with temperature, meaning that the higher the temperature of the exit gas, the greater the loss of energy from unnecessary air escaping through the flue.

BURNING WITH EXCESS FUEL

On the other hand, it is more wasteful to burn with excess fuel, or not enough air. Only the amount of fuel needed to reach a stoichiometric balance with the available oxygen will burn, sending unused fuel up and out the stack. Energy, as well as fuel, is wasted in this situation since the fuel is not burning as efficiently as it potentially could. This also can result in high levels of carbon monoxide in the flue gas. Carbon monoxide in the flue gas can result in soot formation and lower heat transfer effectiveness. For example, if a burner is operated with excess amounts of methane, carbon monoxide and hydrogen will appear as byproducts of the reaction. These combustibles will escape from the process instead of being consumed as they would be if sufficient air with proper mixing had been available. In high-enough concentrations, these combustibles also could create a potentially explosive environment.

CONTROLLING AIR/FUEL RATIO

The best way to achieve complete fuel consumption with low energy waste is to measure the byproducts of the combustion process. The availability of reliable, inexpensive flue gas analyzers insures that combustion may now be controlled with precision. Both excess oxygen and excess fuel can be measured as they leave the burner and stack. Analyzers provide either exact amounts of these elements or percentages of the whole. With these numbers, air dampers can be opened or closed and fuel flows can be adjusted.

MEASURING OXYGEN IN FLUE GASES

There are four methods currently in use to measure oxygen in flue gases. They are the Orsat test, paramagnetic oxygen sensors, wet electrochemical cells and zirconium oxide cells.

Orsat test. One of the earliest methods of measurement, the manually performed Orsat test is still used today. A sample of flue gas, which has been conditioned (cleaned, dried and cooled), is passed through a series of pipets each of which contains a separate chemical reagent. The reagents each absorb a different chemical in the gas—usually oxygen, carbon monoxide and carbon dioxide.

As the gas passes through each pipet, its volume is measured. Any change in measurements indicates the amount of a particular gas that was absorbed.

There are several disadvantages to the Orsat test. It is slow, repetitive work and its accuracy depends on the purity of the reagents and the skill of the operator. Also, there is no way to provide an automatic signal for a recorder or control system.

Paramagnetic oxygen sensor. This sensor takes advantage of the fact that oxygen molecules are strongly influenced by a magnetic field. One common design uses two diamagnetic nitrogen-filled quartz spheres connected by a quartz rod in a dumbbell shape. The dumbbell is supported and suspended in a nonuniform magnetic field. Since the spheres are diamagnetic, they will swing away from the magnetic field.

When a gas containing oxygen is introduced into the spheres, the dumbbell will swing toward the magnetic field across a distance that is proportional to the amount of oxygen in the gas. This movement can be detected either optically or electronically.

Since it is a delicate process, paramagnetic sensors work best in a laboratory and not in an industrial setting. Any sample of flue gas used must be cleaned, dried and cooled before being put into the mechanism. Flue gas constiuents, such as nitrous oxide and some hydrocarbons, have paramagnetic properties that interfere with the test results.

Wet electrochemical cells. These cells use two electrodes in contact with an aqueous electrolyte through which gases containing oxygen are passed. The oxygen in the gas enters into a chemical reaction in which four electrons from each oxygen molecule release hydroxyl ions into the electrolyte at a cathode. These hydroxyl ions in turn react with a lead or cadmium anode with the subsequent release of four electrons to an external circuit. The net flow of electrons creates an electrical current which is proportional to the amount of oxygen passing through the cell.

Wet cells require sample conditioning of flue gas before it can be released into the cell. Without such cooling and cleaning, the cell membrane quickly becomes coated and ceases to function. The cells also must be stored in air-tight containers since any oxygen, not just that from a flue gas sample, will cause the anode to oxidize.

Zirconium oxide cell. Zirconium oxide testing was developed as a byproduct of the U.S. space program. Because of its ability to measure oxygen in hot dirty gases without sample conditioning, it quickly became an industry standard.

The heart of the sensing element is a closed-end tube made of ceramic zirconium oxide stabilized with an oxide of yttrium or calcium. Porous coatings of platinum on the inside and outside serve as electrodes. At high temperatures (normally above 1200 degrees F), oxygen molecules coming in contact with the platinum pick up four electrons and become highly mobile oxygen ions. As long as the concentration of oxygen is equal on each side of the cell, there is no movement of ions through the zirconium.

When the two electrodes are in contract with gases having different oxygen partial pressures, ions move from the area of higher pressure to that of lower pressure, creating a difference in voltage between the electrodes.

When the partial pressure of one gas (usually air) is known, the electrical current created is a measure of the pressure and oxygen content of the other gas. In equation form, the voltage shift is equal to a predetermined constant multiplied by the logarithm of the ratio of two different oxygen partial pressures.

The constant is based on the temperature of the zirconium cell, standard gas laws and free electron values. The cell produces zero voltage when air is on both sides. Under other conditions, this voltage increases as the oxygen concentration of the sample decreases.

One of the key advantages of the zirconium oxide cell is that it operates at high temperatures, which means there is no need to cool or dry the flue gas before it is analyzed. Most zirconium cells make direct measurements in or near the stack with the only protection being a filter to keep ash out of the sampling chamber. Unlike the wet electrochemical cell, the zirconium oxide cell has a virtually unlimited shelf life.

TYPES OF OXYGEN ANALYZERS

Both the paramagnetic and the wet electrochemical cell analysis requires sample conditioning to clean, cool, and dry the flue gas before measurement can be made. The Orsat test cannot

be used for a continuous on-line analysis of flue gas. The zirconium oxide cell is the only continuous analyzing method that can
be performed directly on the stack without the need for sample
conditioning.

Just as there are different methods of determining the amount
of oxygen in a flue gas sample, there also are three different
arrangements by which the sensor units are brought into contact
with gases on the stack to measure oxygen. These types of analyzers are: (1) in situ, (2) convective, (3) close-coupled extractive.
While a fourth type, the extractive analyzer, will work off a long
line from the stack, the samples need to be cooled, cleaned and
dried before they can be tested.

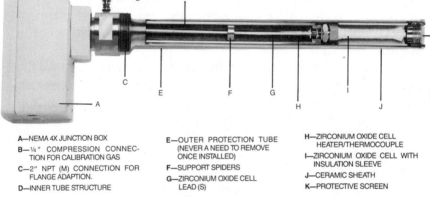

A—NEMA 4X JUNCTION BOX	E—OUTER PROTECTION TUBE	H—ZIRCONIUM OXIDE CELL
B—¼″ COMPRESSION CONNEC- TION FOR CALIBRATION GAS	(NEVER A NEED TO REMOVE ONCE INSTALLED)	HEATER/THERMOCOUPLE I—ZIRCONIUM OXIDE CELL WITH
C—2″ NPT (M) CONNECTION FOR FLANGE ADAPTION.	F—SUPPORT SPIDERS G—ZIRCONIUM OXIDE CELL	INSULATION SLEEVE J—CERAMIC SHEATH
D—INNER TUBE STRUCTURE	LEAD (S)	K—PROTECTIVE SCREEN

Fig. 7-1. Cutaway View of an In Situ Oxygen Analyzer

In situ analyzer. As its name implies, the in situ analyzer is
placed directly in the flow of the flue gas. The zirconium oxide
cell is located at the end of a stainless-steel probe nine inches to
nine feet in length, depending on the application. (See Fig. 7-1)

A heating element, in conjunction with a thermocouple, controls the cell temperature to ensure proper operation. A flame
arrestor can be placed ahead of the cell to prevent the hot zirconium oxide from igniting any combustibles in the stack. Flue
gas diffuses into the probe opening and comes in contact with the
zirconium oxide. The voltage created by the difference in oxygen
pressure is carried by a cable to the control unit where it is changed
to an output signal suitable for an automatic controller or recorder.

The compact design of an in situ analyzer makes it suitable for many industrial applications. With the addition of a filter element, in situ sensors can be used in such dirty testing environments as cement kilns and recovery boilers.

There are some drawbacks in its applications. Since its analyzing units are located directly in the stack, the in situ unit cannot be used in applications where temperatures are more than 1250 degrees F. A convective or close-coupled unit would be more applicable in such circumstances. One other drawback to older in situ models has been difficulty of servicing them. When an in situ probe stopped functioning, it had to be taken completely off line and shipped back to its manufacturer for repairs.

Newer in situ units, however, employ a modular construction and the internal unit, which includes the cell, furnace and thermocouple, can be removed for on-site inspection and repair. Parts can be unscrewed and replaced in minutes, instead of the weeks or months needed for a factory repair.

These newer models also have microprocessor-based controls which make calibration, maintenance and repair easier. (See Fig. 7-2.) An electronic in situ probe can be calibrated with the push of one button, in contrast to the tedious task of hand-adjusting the older analog systems which remain susceptible to fading and drifting.

Maintenance and repair of these newer systems is made even easier by a self-diagnostic system which, through the use of digital codes, indicates what is wrong and what needs to be fixed or replaced.

Convective analyzer (hybrid model). This type of analyzer uses the physical property of convection to move sample flue gas to the zirconium oxide cell located just outside the process wall. Since hot air rises, the oxygen-sensing cell is placed above the level of the gas inlet pipe. As gas in the vicinity of the cell is heated, it rises up and out of the cell housing and is replaced by gas being drawn out of the filter chamber and into the inlet pipe. The gas that has left then cools off on its way back into the filter chamber through a continuous loop. (See Fig. 7-3.) Process gas is constantly diffusing in and out of the filter chamber.

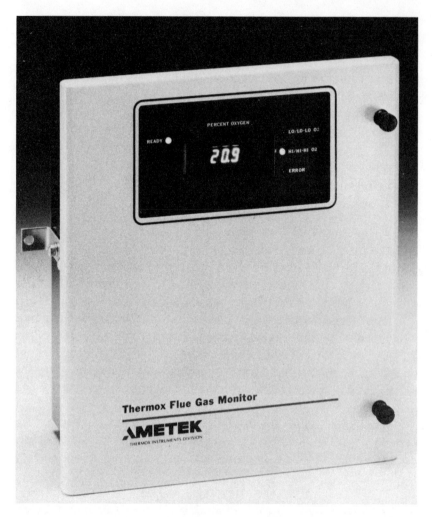

Figure 7-2. Microprocessor-based Control Unit

The temperature between the gases can differ by as much as 1300 degrees F inside the cell housing when passing the zirconium cell and 400 degrees F on the return loop outside it, where the temperature differential sets up the convection flow.

The intake area of the convective analyzer is surrounded by a filter. This makes it ideal for use in such high particulate applications as coal, cement and waste incinerators, and recovery boilers.

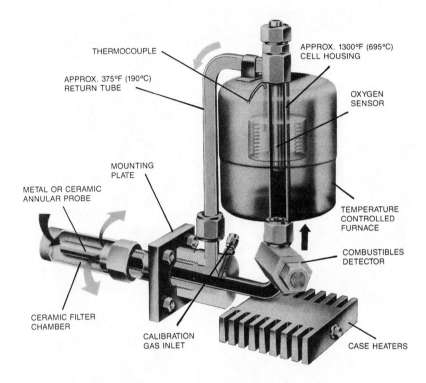

THERMOCOUPLE

APPROX. 1300°F (695°C)
CELL HOUSING

APPROX. 375°F (190°C)
RETURN TUBE

OXYGEN
SENSOR

MOUNTING
PLATE

METAL OR CERAMIC
ANNULAR PROBE

TEMPERATURE
CONTROLLED
FURNACE

COMBUSTIBLES
DETECTOR

CERAMIC FILTER
CHAMBER

CALIBRATION
GAS INLET

CASE HEATERS

Fig. 7-3. Cutaway View of a Convective Analyzer

Since gases diffuse through the filter and are drawn into the analyzer by convection, the force on the process gas is not great enough to pull unwanted particles through the filter and into the cell.

The convective probe can be used in temperatures of up to 2800 degrees F. Its only limitation is the length of the inlet probe, which is a maximum of 48 inches. Like the in situ, newer models of convective units are available with microprocessor-based controllers to help with calibration and maintenance. All working parts are located outside the stack, so most repairs can be done on site.

Close-coupled extractive analyzer. Unlike the in situ and convective probes, a close-coupled extractive probe uses the force of an air-driven aspirator to pull flue gas samples into the analyzer. The sensor is located just outside the process wall and is connected to a probe that protrudes into the flue gas stream. (See Fig. 7-4.)

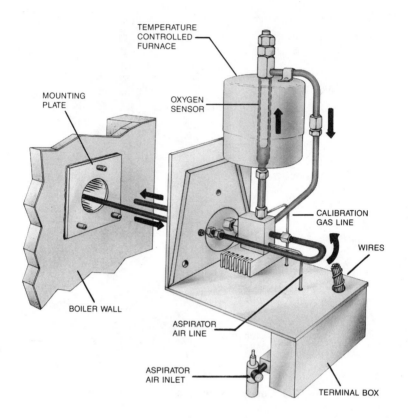

Figure 7-4. Cutaway of a Close-coupled Extractive Analyzer

Flue gas is pulled into the heated sampling area by the aspirator, which creates a vacuum by forcing air out the other end of the loop. The flue gas enters the pipe to fill the vacuum and about 5 percent of it is lifted into the furnace and cell through the same convection process used in the convective analyzer.

Since the sensor is located so close to the stack and is heated, no sample conditioning is needed. As a trade-off for using force to pull samples into the analyzing loop, the close-coupled extractive unit must be used in relatively clean burning applications, such as natural gas and some lighter grades of oil. This type of sensor yields the fastest response to process changes.

There is no practical limit on the length of the probe and the analyzer can be used at temperatures of up to 3200 degrees F. As with the other two analyzers already described, newer models of the close-coupled extractive unit are available with microprocessor-based controllers.

Extractive analyzers. Extractive analyzers are not usually considered to be stack-mounted sensor units. This is because, in a number of units, the gas is being extracted as far as 50 to 100 feet away from the stack for analysis. Once the sample gas reaches the analyzer, it must be conditioned (cooled, cleaned and dried) before being tested by an Orsat, paramagnetic oxygen, wet electrochemical or zirconium oxide sensor.

MEASURING COMBUSTIBLES IN FLUE GASES

There are three methods currently in use to measure such flue gas combustibles as carbon monoxide and hydrogen. They are wet electrochemical cells, catalytic combustibles detectors and non-dispersive infrared absorption.

Wet electrochemical cells. This method for measuring carbon monoxide is very similar to that of electrochemical oxygen detection cells. Carbon monoxide is passed through a membrane, comes into contact with an anode and cathode, and become ionized, creating a voltage difference. This change in voltage is directly proportional to the amount of carbon monoxide that is flowing through the cell.

However, only carbon monoxide, and not hydrogen, can be measured with this type of cell. The cells can become clogged by dirty flue gas, making it necessary to condition samples before analysis.

Catalytic combustibles detector. Catalytic detectors are in wide use to detect combustible gases in the air of such closed areas as mines and parking garages. Recently, versions of these detectors have been put to use in flue gas applications.

In a catalytic sensor, oxygen and combustibles are brought together to burn at temperatures below their normal reaction point of more than 1000 degrees F. This is done by bringing them in contact with a catalyst element. In the case of flue gas, this catalyst usually is platinum, which is maintained at a temperature of about 400 degrees F.

The detector itself consists of two resistance elements protruding into the sample stream of flue gas. One of the elements, the active element, is coated with the platinum catalyst while the other element, the reference element, is coated with an inert binder.

When it is heated to more than 400 degrees F, the sensor causes any oxygen and combustibles in the flue gas to burn when they come in contact with the platinum. This adds heat to the active element and changes its electrical resistance. The difference between the resistance of the two elements is directly proportional to the amount of combustibles in the flue gas sample passing through the sensor.

Quality flue gas combustibles analyzers with full-scale ranges, as sensitive as 0 to 2000 parts per million, use thin film resistance elements that are laser trimmed and computer matched for high accuracy. Certain models also are available with microprocessors for added accuarcy and convenience.

Several models of convective and close-coupled extractive oxygen analyzers incorporate catalytic combustible sensors, providing both oxygen and combustible measurements from the same unit.

Non-dispersive infrared absorption. Carbon monoxide is one of many gases known to absorb infrared energy of specific wavelengths. Use of this principle allows the presence of carbon monoxide to be measured using light waves. If a beam of infrared light is passed through a gas sample containing carbon monoxide and if the emerging energy is measured with a spectrophotomer, significantly less energy will be detected for wavelengths between 4.5 and 4.7 microns than if the carbon monoxide were not present.

The amount of energy absorbed is a measure of the concentration of carbon monoxide in the sample through the use of Beer's Law. If the length the infrared energy travels is kept constant, the amount of carbon monoxide present is directly proportional to the ratio of the infra-red intensity before passing through the gas and its intensity after passing through it.

This infrared absorption method can be used to identify carbon monoxide in two types of analyzers—off-stack and across-the-stack.

As its name imples, an off-stack absorption analyzer tests flue gas for carbon monoxide after it is extracted from the stream in the stack. The gas is cooled, cleaned, dried and placed in a sample cell. Bursts of infrared energy are shot through the cell and through a reference cell containing air. These bursts then pass through an optical filter that removes all wavelengths but those between 4.5 and 4.7 microns. The difference between the energy in the two paths indicates how much carbon monoxide is in the flue gas sample.

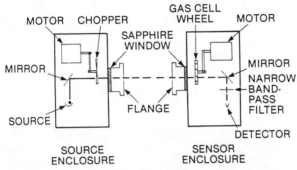

Fig. 7-5. Typical Across-the stack CO Analyzer

Unlike the off-stack sensor, which reads the carbon monoxide concentration at the point the sample was taken, an in situ across-the-stack unit measures the amount of carbon monoxide in a cross-section of the entire flue gas stream.

An across-the-stack analyzer shoots its bursts of infrared energy directly through the flue. (See Figure 7-5.) Instead of passing through a reference tube of air after reaching the other side, the bursts pass through two rotating cells. One of these cells contains pure carbon monoxide, which blocks out all the 4.5-4.7 micron radiation. The other one contains pure nitrogen, which allows all the waves to pass through. The difference in these signals is inversely proportional to the carbon monoxide present in the stack.

Besides providing full-stream readings, the across-the-stack analyzer offers an instantaneous response to changes in carbon monoxide concentrations. Newer models are lightweight, much more compact, and can be purchased with microprocessor controls.

CHOOSING THE RIGHT ANALYZING SYSTEM

The choice of a flue gas analyzing system must be based on the specific needs of the application. For example, the operator of a resource recovery burner does not need to be as concerned about fuel efficiency—since the fuel is sludge or garbage—as about the amount of carbon monoxide being released into the air. On the other hand, the operator of an electric utility, which burns pulverized coal, must be concerned with measuring both oxygen and combustibles in a dirty stack environment.

Deciding the exact needs of a process is the first step in choosing a system of sensors. Once that is done, there are a number of analyzers that could work for specific situations. Combination oxygen-combustible analyzers, with microprocessor controls, are among the best in design, but the lowest in cost. (See Fig. 7-6.) There are systems made of a combination of in situ oxygen analyzers and infrared carbon monoxide analyzers for special applications.

Fig. 7-6. Oxygen and Combustibles Control Unit

When shopping for analyzers, though, it is important to make comparisons on the basis of the same specifications. Some systems which at first glance seem to meet all expectations, may upon closer inspection not be based in a true application. So before making the final selection, know the application, consult with the manufacturers and study the specifications to make sure that the right analyzer is being purchased for the job.

8

Boiler Test Procedures

The principal method used for improving boiler efficiency involves operating the boiler at the lowest practical excess O_2 level with an adequate margin for fuel variations in fuel properties and ambient conditions and the repeatability and response characteristics of the combustion control system. These tests should only be conducted with a thorough understanding of the test objectives.

BOILER TEST PROCEDURES

Cautions

- Extremely low excess O_2 operation can result in catastrophic results.

- Know at all times the impact of the modification on fuel flow, air flow and the control system.

- Observe boiler instrumentation, stack and flame conditions while making any changes.

- When in doubt, consult the boiler manufacturer.

- Consult the boiler operation and maintenance manual supplied with the unit for details on the combustion control system or methods of varying burner excess air.

Test Description

The test series begins with baseline tests that document existing "as-found conditions" for several firing rates over the boiler's normal operating range. At each of these firing rates, variations in excess O_2 level from 1 to 2% above the normal operating point to the "minimum O_2" level are made. Typical ranges of minimum excess O_2's are given on Table 8-1.

TABLE 8-1. Typical Excess Air Requirements and Resultant O_2
In Flue Gas at Furnace Outlet

	Excess Air (%)	O_2 in Flue Gas (%)
Coal		
Cyclone furnace	7-15	1.5-3
Pulverized firing	20-30	4-5
Stoker firing	25-40	4.5-6.5
Oil	3-15	<1-3
Natural Gas	5-10	1-2

Curves of combustibles as a function of excess O_2 level will be constructed similar to those given in Figures 8-1 and 8-2. As illustrated in these figures, high levels of smoke or CO indicating potentially unstable operation can occur with small changes in excess O_2 so that small changes in excess O_2 should be made for conditions near the smoke or CO limit.

It is important to note that the boiler may exhibit a gradual smoke or CO behavior at one firing rate and a steep behavior at another. Minimum excess O_2 will be that at which the boiler just starts to smoke, or the CO emissions rise above 400 ppm or the Smoke Spot Number equals the maximum value as given in Table 8-2.

Once minimum excess O_2 levels are established, an appropriate O_2 margin or operating cushion ranging from 0.5 to 2.0% O_2 above the minimum point is set, depending on the particular boiler control system and fuels.

Repeated tests at the same firing condition approaching from both the "high side" and "low side" (i.e., from higher and lower firing rates) can determine whether there is excessive play in the boiler controls. Record all pertinent data for future comparisons. Readings should be made only after steady boiler conditions are reached and at normal steam operating conditions.

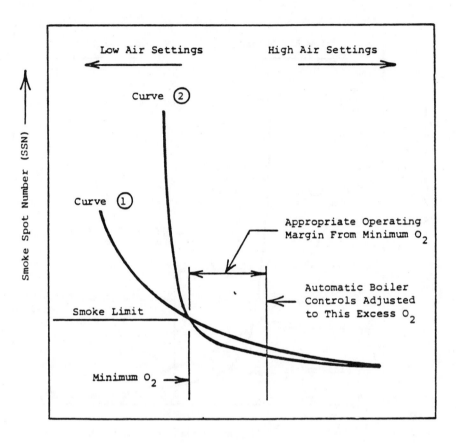

Curve 1 - Gradual smoke/O_2 characteristic
Curve 2 - Steep smoke/O_2 characteristic

FIGURE 8-1. Typical smoke-O_2 characteristic curves for coal- or oil-fired industrial boilers.

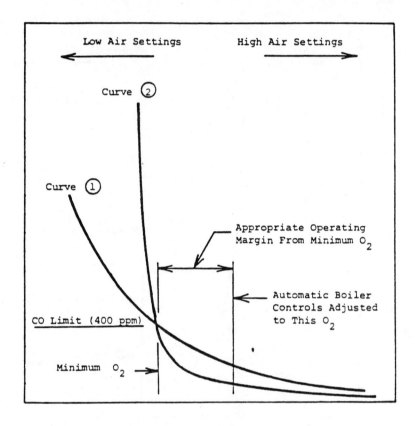

Percent O_2 in flue gas ⟶

Curve 1 - Gradual CO/O_2 characteristic
Curve 2 - Steep CO/O_2 characteristic

FIGURE 8-2. Typical CO-O characteristic curve for gas-fired industrial boilers.

TABLE 8-2. Maximum Desirable Smoke Spot Number

Fuel Grade	Maximum Desirable SSN
No. 2	less than 1
No. 4	2
No. 5 (light and heavy), and low-sulfur resid	3
No. 6	4

Step-By-Step Boiler Adjustment
Procedure for Low Excess O_2 Operation

1. Bring the boiler to the test firing rate and put the combustion controls on manual.

2. After stabilizing, observe flame conditions and take complete set of readings.

3. Raise excess O_2 1-2%, allowing time to stabilize and take readings.

4. Reduce excess O_2 in small steps while observing stack and flame conditions. Allow the unit to stabilize following each change and record data.

5. Continue to reduce excess air until a minimum excess O_2 condition is reached.

6. Plot data similar to Figures 8-1 and 8-2.

7. Compare the minimum excess O_2 value to the expected value provided by the boiler manufacturers. High excess O_2 levels should be investigated.

8. Establish the margin in excess O_2 above the minimum and reset the burner controls to maintain this level.

9. Repeat Steps 1-8 for each firing rate to be considered. Some compromise in optimum O_2 settings may be necessary since control adjustments at one firing rate may affect conditions at other firing rates.

10. After these adjustments have been completed, verify the operation of these settings by making rapid load pick-ups and drops. If undesirable conditions are encountered, reset controls.

11. Low fire conditions may require special consideration.

12. Perform tests on any alternate fuel used. Again, some discretion will be required to resolve differences in control settings for optimum conditions on the two fuels.

Evaluation of The New Low O_2 Settings

Extra attention should be paid to furnace and flame patterns for the first month or two following implementation of the new operating modes. Thoroughly inspect the boiler during the next shutdown. To assure high boiler efficiency, periodically make performance evaluations and compare with the results obtained during the test program.

BURNER ADJUSTMENTS

Adjustments to burner and fuel systems can also be made in addition to the low excess O_2 test program previously described. The approach in testing these adjustments is a "trial-and-error" procedure with sufficient organization to allow meaningful comparisons with established data.

Items that may result in lower minimum excess O_2 levels include changes in:

- burner register settings
- fuel oil temperature
- oil gun tip position
- fuel and atomizing pressure
- oil gun diffuser position
- coal particle size
- coal spreader position

Evaluation of each of these items involves the same general procedures, precautions and data evaluation as outlined previously. The effect of these adjustments on minimum O_2 are variable from boiler to boiler and difficult to predict.

9

Efficiency-Related Boiler Maintenance Procedures

This chapter can assist the purchasers and operators of industrial and commercial boilers in establishing maintenance programs specifically aimed at maintaining high efficiency levels for gas-, oil- or coal-fired units in sizes ranging from 10,000 to 500,000 lb/hr steam flow.

Efficiency-related boiler maintenance aims to correct any condition, other than a fuel change, which results in increased energy consumption to generate a given amount of steam.

Thus, at constant boiler load any condition is considered efficiency-related if the component malfunction or performance degradation resulted in an increase in:

1. stack gas temperature
2. excess air requirements
3. carbon monoxide, smoke or unburned carbon in ash
4. convection or radiation losses from the boiler exterior, ductwork and piping
5. blowdown above that required to maintain permissible boiler water concentration
6. auxiliary power consumption by fans, pumps or pulverizers.

This definition distinguishes efficiency-related maintenance items from maintenance normally performed for personnel or equipment protection. The maintenance procedures discussed in this chapter are directed to maintaining efficient operations and are not intended to replace any procedures which are required for safety reasons or recommended by the manufacturer to extend equipment life.

In fact, efficiency-related boiler maintenance might be looked upon simply as a more intensive application of preventive maintenance procedures in areas where efficiency is affected before the equipment has deteriorated to the point where reliability, safety or capacity is threatened. For example, ash or soot deposits on heating surfaces may reduce heat transfer and increase exit flue gas temperature but would not affect plant safety or load carrying capability unless the deposits restricted gas passages to the extent that insufficient air could be supplied to the unit.

EFFICIENCY SPOTCHECK

A spotcheck of combustion conditions using stack measurements of O_2, CO, smoke and temperature as well as flame appearance is an effective maintenance tool. These tests should be conducted at operating conditions corresponding to those used in establishing performance goals.

Again, test data should be recorded so that deterioration in operating efficiency can be readily identified. The frequency of these spotchecks will depend on the complexity of the system and manpower available but should be on at least a weekly basis.

ESTABLISHING PERFORMANCE GOALS

The initial requirement of an efficiency-related boiler maintenance program is to establish best achievable performance of the unit over the operating load range.

For new units, data taken during initial startup should be compared to manufacturer's predicted performance to establish that the unit is operating as designed. Every effort should be made to resolve performance discrepancies while the manufacturer's service representative is on the job. Some increase in gas temperatures can be expected in the first few months of operation on oil- or coal-fired units as gas side deposits stabilize.

On older units, baseline data should be taken over the load range after the boiler has had an extensive outage for cleaning, inspection and repair followed by a control system and burner system tune-up. With effective cleaning, repair and tune-up it should be possible to operate at full load with 15% or lower excess air at the boiler outlet on gas firing, 20% or lower excess air on oil firing, 20 to 30% excess air on pulverized coal firing and 30 to 40% excess air on stoker firing.

The exit gas temperature on firetube or watertube boilers without economizers or air heaters should not be greater than 200°F above saturated steam temperature at full load for the majority of designs. Less favorable excess air or exit gas temperature would indicate the potential for further improvement in operation.

When air heaters and/or economizers are installed, variations in design make generalization on exit gas temperature impossible. Gas and air side pressure drops are also extremely design dependent and defy generalization. Design data is therefore necessary to establish expected performance. The boiler manufacturer should be able to supply this information if plant records are incomplete.

PERFORMANCE MONITORING (Boiler Log)

The objective of performance monitoring is to document deviations from desired performance as a function of time. This information should be obtained on a regular basis by the operator under steady load conditions. Data taken during load changes or under fluctuating load conditions will be inconsistent and of little value in assessing unit efficiency; however, control response should be noted during transient conditions.

If widely fluctuating load is the normal condition, it may be necessary to make special arrangements to achieve steady boiler load for efficiency monitoring either through curtailing intermittent steam demands or taking load swings on other boilers. If the boiler master control is placed on manual operation the fuel/air ratio should be as established by the control system.

The performance recorded under these conditions will indicate deviation from desired fuel/air ratio and other performance deviations. By manually adjusting fuel/air ratio to the desired level a second set of data can be obtained which represents performance deviations attributable to sources other than the fuel/air ratio such as surface cleanliness, boiler baffles, etc.

The actual readings to be taken and the frequency are determined by the size and complexity of the equipment and the manpower which can be justified in collecting and analyzing data. Usual practice is to record data hourly to check general performance. These hourly readings are to assure that the unit is operating normally and includes checking of safety and mechanical devices.

Efficiency-related items which should be included in the boiler operator's log are:

1. General data to establish unit output
        steam flow, pressure
        superheated steam temperature (if applicable)
        feedwater temperature

2. Firing system data
        fuel type (in multi-fuel boilers)
        fuel flow rate
        oil or gas supply pressure
        pressure at burners
        fuel temperature
        burner damper settings
        windbox to furnace air-pressure differential
        other special system data unique to particular installation

3. Air flow indication
        air preheater inlet gas O_2
        stack gas O_2
        optional - air flow pen, forced draft
            fan damper position, forced draft
            fan amperes

4. Flue gas and air temperatures
        boiler outlet gas
        economizer or air heater outlet gas
        air temperature to air heater

5. Unburned combustible indication
        CO measurement
        stack appearance
        flame appearance

6. Air and flue gas pressures
   forced draft fan discharge
   furnace
   boiler outlet
   economizer differential
   air heater air and gas side differencial

7. Unusual conditions
   steam leaks
   abnormal vibration or noise
   equipment malfunctions
   excessive makeup water

8. Blowdown operation

9. Sootblower operation.

While this list may look extensive and time consuming the operator of a firetube or comparably sized watertube boiler (10,000-24,000 lb steam per hour) will find that the data list will reduce to:

steam pressure
feedwater temperature
steam, feedwater or fuel flow
fuel supply pressure
fuel supply temperature
boiler outlet gas temperature
boiler outlet O_2
F.D. fan inlet temperature
stack appearance
flame appearance
windbox air pressure
windbox to furnace air-pressure differential
boiler outlet flue gas pressure
blowdown operation
unusual conditions or equipment malfunctions.

If the unit is too small to justify a continuous O_2 analyzer, the excess air should be checked by Orsat analysis weekly. If the burner is serviced by an outside organization the frequency should be at least monthly, and include an excess air determination. Carbon monoxide determination is particularly important on gas firing since high CO levels can develop without smoke formation, unlike oil or coal firing.

If steady load conditions are difficult to obtain, less emphasis should be placed on the boiler operator's log for efficiency-related maintenance items and regularly scheduled performance checks should be made under steady conditions on a monthly basis for smaller units and on a bi-weekly basis for larger units. On coal-fired units monthly checks on ash combustible content should be made. On pulverized coal-fired units coal fineness should also be checked monthly.

PERIODIC EQUIPMENT INSPECTION

Items included in the Preliminary Boiler Inspection Checklist (Table 6-1) should be used as a basis for the periodic equipment inspection. Note that several of these items are included in the performance monitoring log and will therefore be inspected on a frequent basis. Other conditions should be checked during annual outages or more frequently if possible.

PERFORMANCE TROUBLESHOOTING

Specific deviations from expected performance can be of considerable value in determining the cause of performance deficiency. Table 9-1 presents a summary of frequent problems encountered in boiler systems and the possible causes.

TABLE 9-1. Boiler Performance Troubleshooting

System	Problem	Possible Cause
Heat Transfer Related	High exit gas temperatures	– Buildup of gas- or water-side deposits
		– Improper water treatment procedures
		– Improper sootblower operation
Combustion Related	High Excess air	– Improper control system operation
		– Low fuel supply pressure
		– Change in fuel heating value
		– Change in oil fuel viscosity
	Low excess air	– Improper control system operation
		– Fan limitations
		– Increased ambient air temperature
	High CO and combustible emissions	– Plugged gas burners
		– Unbalanced fuel/air distribution in multi-burner furnaces
		– Improper air register settings
		– Deterioration of burner throat refractory
		– Stoker grate condition
		– Stoker fuel distribution orientation
		– Improper overfire air systems
		– Low fineness on pulverized systems

(continued)

System	Problem	Possible Cause
Miscellaneous	Casing leakage	— Damaged casing and insulation
	Air heater leakage	— Worn or improperly adjusted seals on rotary heaters
		— Tube corrosion
	Coal pulverizing power	— Pulverizer in poor repair
		— Too low classifier setting
	Excessive blowdown	— Improper operation
	Steam leaks	— Holes in waterwall tubes
		— Valve packing
	Missing or loose insulation	— Overheating
		— Weathering
	Excessive sootblower operation	— Arbitrary operation schedule that is in excess of requirements

(end)

PERFORMANCE DEFICIENCY COSTS

The benefits by maintaining high boiler efficiency are readily calculated knowing boiler fuel usage (W_f), efficiency achievable (E_A), efficiency differential between achievable and actual performance (ΔE), and fuel cost, (C).

$$\text{fuel \$ saving} = W_f \ x \ \frac{\Delta E}{E_A} \ x \ C$$

The units selected must be consistent, that is, if W_f equals million Btu/year, and C equals dollars per million Btu, the fuel saving will be in dollars per year.

Usually, achievable efficiency and ΔE will vary with boiler load. To evaluate annual fuel savings under varying load conditions the fraction of operating time spent at each load must be estimated and multiplied by the fuel consumption rate and the ratio of $\Delta E/E_A$ for each load condition. The sum of the individual load calculations multiplied by fuel cost will then equal the annual fuel dollar saving.

To provide some estimate as to the value of a one percent efficiency change, consider a boiler operating at 10,000 lb per hour steam flow for 6,000 hours per year. The heat required to generate one pound of steam is approximately 1000 Btu divided by boiler efficiency as a decimal fraction. If efficiency is 80% the annual fuel consumption would be

$$10{,}000 \ x \ 6{,}000 \ x \ \frac{1{,}000}{.80} = 75{,}000 \text{ million Btu/year}$$

With fuel costs at $6 per million Btu, the annual fuel bill would be $450,000.

If efficiency decreased to 79% the fuel bill would be $455,696 or $5,696 higher.

If the boiler steam flow were 100,000 lb/hr and the efficiency decrease was 2% the increase in annual fuel cost would be $114,000. Obviously, more effort can be justified in maintaining efficiency as boiler output increases and performance deterioration becomes greater.

Efficiency gains through reducing excess air or stack temperature are shown in Figures 9-1 and 9-2. A reduction in excess air is usually accompanied by a reduction in stack temperature. The actual temperature reduction is dependent on the initial stack temperature and the excess air reduction. A 20% excess air reduction will produce approximately a 30°F reduction if exit gas temperature is in the 500-600°F range, but only a 15°F reduction if exit gas temperature is in the vicinity of 300°F.

The combined effect can be determined by first evaluating the efficiency improvement for reduced excess air from Figure 8-1 and then adding the efficiency improvement for reduced stack gas temperature from Figure 9-2 evaluated at the lower excess air.

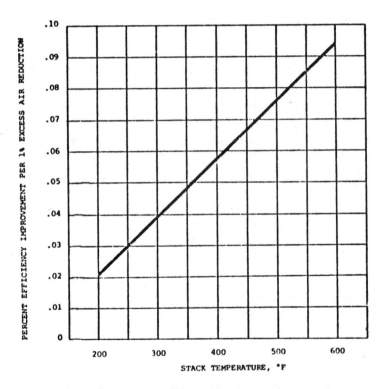

FIGURE 9-1. Percent efficiency improvement for every one percent reduction in excess air. Valid for estimating efficiency improvements on typical natural gas, #2 through #6 oils and coal fuels.

Solid combustible losses are of concern principally with coal firing as discussed in Chapter 3. Objectionable smoke will occur with oil or gas firing before unburned carbon losses become significant. In a given coal-fired installation, solid combustible losses are strongly influenced by excess air, tending to increase as excess air is decreased. Best operation involves determining the excess air at which *total* heat losses are minimized.

Carbon monoxide heat losses can become significant with gas or oil firing. Carbon monoxide in the flue gas at a level of 1000 ppm or 0.1% represents a heat loss of approximately 0.35%. Higher CO levels will produce correspondingly higher heat losses.

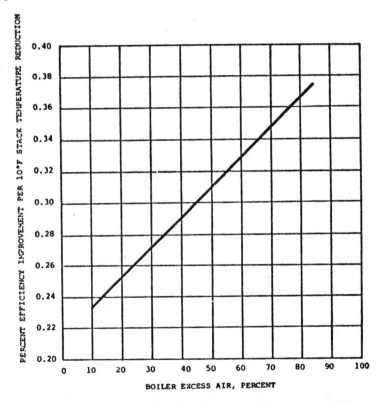

FIGURE 9-2. Percent efficiency improvement for every 10°F drop in stack temperature. Valid for estimating efficiency improvements on typical natural gas, #2 through #6 oils and coal fuels.

10

Boiler Tune-Up

One of the more effective means of improving and maintaining boiler operating efficiency is a boiler "tune-up." This preventive maintenance item is one of the most direct approaches to fuel conservation through efficiency improvements.

The primary objective in a "tune-up" is to achieve efficient combustion with a controlled amount of excess air. Operating with the lowest practical excess air will minimize efficiency losses by reducing the quantity of unneeded air which is heated to the stack temperature then not utilized. The associated reduction in stack gas temperature and power consumption by forced draft and induced draft fans are additional benefits.

The actual improvement in boiler efficiency with lower excess air depends on the initial stack temperature and excess air at the boiler exit. A given change in excess air will have a greater effect when stack temperatures are high. For example, a reduction in excess air of 10% (i.e., from 20 to 10% or 100 to 90% excess air) will produce a 0.9% improvement in efficiency when stack temperatures are on the order of 600°F.

However, the same change in excess air when stack temperatures are near 200°F will improve boiler efficiency by only 0.2%. These values are not appreciably affected by the type of fuel (natural gas, oils or coal). Cases have been documented where reductions in excess air by much more than 10% were possible leading to efficiency improvement of several percent.

Proper operation of the boiler combustion control system is essential for maintaining high boiler operating efficiencies and satisfactory excess air levels. Its main purpose is to provide the correct quantities of fuel and air at the burner to satisfy a varying demand for steam generation.

Although it is important that the excess air delivered to the burner be kept to a minimum over the boiler operating range, it generally is not practical to operate precisely at the point of maximum efficiency. This optimum typically occurs at the "threshold" of combustible or smoke formation and may result in unacceptable stack conditions.

For the majority of boilers, it is necessary to maintain a margin of excess air above the minimum or threshold level to accommodate variations in fuel properties and ambient conditions, non-repeatability of control settings, normal deterioration of control parts and rapid changes in firing rate.

Proper adjustment and maintenance of the burning equipment are essential to good combustion control. Burner parameters such as air register position, location of the diffuser or position of coal spreaders can affect the excess air requirements. Fuel oil temperature and atomizing pressures, pulverizer performance (coal fineness), primary air temperature, and other operating factors also have an impact.

To assure reliable, safe, and efficient boiler performance, manufacturers of boiler and burner equipment recommend periodic inspections and tune-ups. Thorough tune-ups are commonly recommended to be performed on an annual basis but many operators also prefer to conduct quick boiler efficiency checks much more often, sometimes on a daily or weekly basis. In this way, efficiency problems can be detected early, before large fuel wastage occurs or expensive maintenance is required.

Boiler tune-up services are available from most major manufacturers of boiler and burner systems, some local utilities, and engineering consulting firms. Since tune-up crews usually charge an hourly or daily rate plus expenses, costs for these services will depend on the extent of the tune-up work required.

A minimal tune-up should include a verification of automatic fuel and air control operation over their operating range. Visual furnace observation and stack measurements of excess O_2, CO, CO_2 and temperature are essential elements in this type of tune-up. A typical tune-up performed on a 13,000 lb/hr watertube boiler during a field testing program could be done in one day.

A more extensive tune-up might also involve a boiler shutdown and a thorough inspection of burner parts, dampers, fuel valves and regulators, refractory, furnace tubes and instrumentation. In many cases the more critical maintenance, instrument calibration and repairs can be made on-site during a brief shutdown.

Major repairs may require an extended shutdown at a later, more convenient date. A list of additional items requiring attention at that time should be furnished. For boiler operators desiring a basic efficiency check of their boiler, the local natural gas supply company may provide this service at little or no charge to customers and may also offer some assistance in adjusting burner controls for peak efficiency.

For a complete combustion control tune-up and inspection, the tune-up crew will prefer to have total control of boiler firing rate to verify control settings over the full turn-down range. Be prepared to accomodate the fluctuations in steam flow and/or pressure. In some cases it may be necessary to install a baffled steam discharge stack when these fluctuations cannot be tolerated in the steam supply system or plant processes.

When stack measurements are being made (O_2, CO, CO_2 and temperature), it is important to request a verification of stack gas uniformity (i.e., a traverse) to assure representative readings. Flue gas samples are usually extracted by means of a single point probe and any stratifications or gradients in the flue gas ducting can lead to erroneous measurements.

It is important that excess air not be reduced at the expense of excessive combustibles (unburned fuel, carbon carryover, CO, etc.) since these can represent significant efficiency losses. More than 400 ppm carbon monoxide (CO) in the stack gases is generally not acceptable.

Finally, it may be worthwhile for future reference to inquire about the margin in excess air used by the tune-up crew to accommodate the various uncontrollable factors mentioned previously.

For specialized combustion or operating problems not routinely solved by plant operating or engineering personnel, assistance from outside consulting or engineering firms should also be considered. Some firms have extensive experience in specific problem areas such as tube corrosion and wastage, excessive carbon carryover, precipitator and mill fires, water quality control, instrumentation malfunctions, and air pollution restrictions.

BOILER TUBE CLEANLINESS

Tube deposits and fouling on the external tube surfaces of a watertube boiler or similar gas side conditions in the gas tubes of a firetube boiler will inhibit the absorption of heat in the boiler and lead to lower efficiencies. This condition will be reflected in high flue gas temperatures when compared to "clean" conditions at a similar firing rate and boiler excess air. The resulting loss in boiler efficiency can be closely estimated on the basis that a 1% efficiency loss occurs with every 40°F increase in stack temperature.

It should be mentioned that water side deposits resulting from inadequate water treatment would also eventually lead to higher stack temperatures and lower efficiencies. However, tube failures due to overheating would generally occur before any substantial efficiency losses are evident.

Poor firing conditions at the burner can be the major cause of gas side tube deposit problems. Insufficient burner excess air or improper burner adjustment and maintenance can lead to excessive carbon and soot formation in the furnace which builds up on furnace wall tubes and convective tube banks.

Other factors to be considered might be fuel oil burning temperature, coal ash fouling properties, pulverizer performance and primary air temperature. A boiler inspection will generally indicate whether soot blowers are performing as designed and are properly located for effective tube cleaning. If existing soot blowers are functioning properly but are inadequate for present fuels, repositioning of soot blowers and/or the installation of new soot blower designs might be considered (see Chapter 14).

Stack temperature measurements are an easy and effective means for monitoring boiler tube cleanliness conditions. Stack temperatures sould be periodically compared to values obtained during start-up or following a boiler tube wash to determine any deviations from "clean" baseline temperatures.

Since stack temperature usually increases with firing rate and excess air, these comparisons should be performed at similar boiler operating conditions. In the absence of any reference temperatures, it is normally expected that stack temperatures will be about 150 to 200°F above the saturated steam temperature at high firing rates in a saturated steam boiler.

Naturally, this does not apply to boilers equipped with economizers or air preheaters. The boiler manufacturer should be able to supply a normal range of stack temperatures for the particular boiler design. When higher stack temperatures are measured, a boiler tube cleaning may be warranted.

Some boiler operators perform tube washes in-house but cleaning services are also available from commercial boiler repair and service companies. These companies also perform water side inspection and cleaning. Various cleaning methods are used including high pressure water, air and steam jets to remove gas side deposits.

Mechanical "shakers" and "knockers" are also used which vibrate off or dislodge the tube deposits for easy removal. The appropriate type of cleaning device will be dictated by the severity and composition of deposits, available access to tube surfaces, boiler construction (firetubes versus watertubes), age of the boiler, and other factors.

The cost of an extensive tube cleaning will typically range from a few hundred dollars for a 10,000 lb/hr boiler to several thousand dollars for larger units. This expense is usually very easy to justify due to the quickly realized savings in fuel costs.

DETERMINING EFFICIENCY-
RELATED MAINTENANCE REQUIREMENTS

The implementation of an efficiency-related maintenance program requires that performance deviations from optimum conditions be monitored and corrected either by operating or maintenance procedures. Many industrial boilers will require additional instrumentation for this purpose. A minimum instrumentation list for monitoring unit performance is given in Table 16-2, Chapter 16.

A basic tool for determining the need for maintenance to restore efficiency is the boiler operator's log shown in Chapter 8. Daily trends in unit performance can be used to determine the deviation from design or optimum performance and establish the frequency for maintenance procedures.

Comparison of daily boiler log data to data obtained with the unit in good operating condition may disclose the need for specific maintenance procedures. These observations should be made under very steady firing conditions. A burner and control system tune-up would be indicated if the unit smokes at previously acceptable O_2 levels. Higher than normal O_2 levels may indicate the need for control system adjustment or in the case of balanced draft units, may indicate excessive boiler setting leakage. A deficiency in air flow capacity may indicate excessive air heater seal leakage, plugged air heaters or other boiler and damper problems. The boiler operator's log will assist in assessing the problem.

Annual outages are also necessary for inspection of areas not accessible during operation to determine condition of boiler baffles, boiler cleanliness, air heater corrosion or seal condition, fan blade condition and damper operation.

The major areas of concern are burner-control system maintenance and boiler cleanliness, as discussed above. Some less common, but significant maintenance items from an efficiency standpoint are discussed next.

SPECIAL MAINTENANCE ITEMS –
COAL-FIRED UNITS

Many spreader stoker fired units incorporate fly ash re-injection from boiler hoppers or mechanical collectors to reduce carbon losses by reburning. Efficiency improvements of up to 5% are obtained by re-injection. Maintenance on this system is required to assure that the mechanical collector tubes are operating efficiently (no holes worn through collecting devices), the hoppers are being emptied and the conveying pipes are not eroded to the point of failure.

Stoker grate surfaces must be maintained to provide good air distribution and minimize sifting losses. Feeder devices must be kept in good repair to provide uniform fuel distribution.

In pulverized coal units the coal fineness should be checked regularly and the mills adjusted to provide the correct fineness. High carbon losses will result with poor grinding. Fans or exhausters handling pulverized coal/air mixtures are subject to erosion which can lead to reduced air flow which reduces mill capacity and can lead to fuel line plugging. Fan blades and grinding elements should be replaced as required to maintain performance. The frequency of replacement will depend on the abrasiveness of the coal being pulverized.

11

Boiler Operational Modifications

This chapter outlines the more promising changes in boiler operating practices that can be employed to reduce fuel consumption. These include not only techniques that improve boiler efficiency but also fuel saving practices such as load management and reduced continuous blowdown practices.

REDUCED BOILER STEAM PRESSURES

Reducing the boiler steam pressure is not generally viewed as an efficiency improvement technique but at those boiler installations where reductions in steam pressure are practical, this can be an effective means for saving up to 1 or 2 percent on the fuel bill. A portion of this savings results from lower stack temperatures and the accompanying increase in boiler efficiency.

Lower steam pressures give lower saturated steam temperatures and in cases without stack heat recovery, a similar reduction in final flue gas temperature will result. The actual reduction in temperature for a given drop in steam pressure will depend on the particular pressure levels involved, since the relationship between saturated steam pressure and steam temperature is not linear (consult the steam tables).

For example, a 50 psi drop from 150 psig to 100 psig will lower the steam temperature (and stack temperature) by 28°F. By comparison, a 50 psi drop from 400 psig will reduce the stack temperature by 12°F. The corresponding boiler efficiency improvements would be approximately .7% and .3% respectively.

There can be other immediate benefits from operating at reduced pressures which lead to lower fuel consumption. These can include lower radiated heat loss from steam mains, less leakage at flanges and packing glands, reduced boiler-feed-pump energy consumption and less energy dissipation at pressure reducing stations.

To successfully operate at reduced pressures, the primary pressure must still be sufficiently high to deliver adequate steam to the various points of end use. Lower pressures may also result in undesirable moisture carryover from the steam drum. The boiler manufacturer may be able to supply improved drum intervals if carryover is a problem.

While these are usually the primary concerns, there may be other important considerations. For example, operating with lower main steam pressure might place tighter limits on acceptable steam pressure fluctuations. Since pressure drops across regulators are less, smaller transients in input pressure are acceptable to assure uniform output. Flow metering devices may also have to be recalibrated or corrected for the lower pressures.

WATER QUALITY CONTROL – BLOWDOWN

Water treatment is an important aspect of boiler operation which can affect efficiency or result in plant damage if neglected. Boiler feedwater contains impurities in solution and suspension.

These impurities concentrate in the boiler water since the steam generated is essentially pure. Proper internal boiler water treatment removes all or part of the scale-forming calcium and magnesium salts from solution.

If these suspended solids are allowed to concentrate beyond certain limits, a deposit or "scale" will form on the boiler heating surfaces which will retard heat transfer and increase tube metal temperatures. This can lead to increased stack gas temperatures which reduces boiler efficiencies.

Even more important may be the probability of furnace tube failures from overheating as a result of the insulating effect of water-side scale. High solids can also interfere with proper operation of the steam separating apparatus in the drum causing boiler water carryover with the steam. If a superheater is installed, the solids will either be deposited on the surface causing metal over-heating or be carried along with the superheated steam. Particulates in the steam can cause further problems in erosion or deposits in turbines and valves.

The concentration of suspended and dissolved solids in boiler water is controlled by removing some of the high-solids boiler water and replacing it with low-solids feedwater, effecting a general lowering of solids concentration in the boiler. This process, which is known as blowdown, can be either intermittent bottom blowdown or continuous blowdown.

Bottom or sludge blowdown is necessary to remove any sludge accumulating in the lowest parts of the boiler system. Continuous blowdown is taken from the point of highest solids concentration usually from the upper drum.

Intermittent bottom blowdown may be sufficient if the feedwater is exceptionally pure such as with a high percentage of returned condensate or evaporated makeup water. Intermittent blowdown is performed manually and therefore may result in wide fluctuations in blowdown patterns as shown in Figure 11-1. Use of frequent short blows as opposed to infrequent lengthy blows is preferred due to reduced treated water losses and lost sensible heat energy in the waste water. The expected fuel savings using this technique are dependent on the existing intermittent blowdown patterns.

Continuous blowdown from the upper drum can reduce the required intermittent bottom blowdown and allow the use of heat recovery devices.

Many applications involve a high percentage of makeup water with varying amounts of impurities. Blowdown rate, which is defined as the ratio of water blown from the boiler to the quantity of boiler feedwater, may range up to 10% depending on the ratio of impurities in the feedwater to allowable concentration in the boiler water.

For instance, the permissible concentration of total solids in boiler water is 3500 ppm at 300 psi or lower operating pressure. If the feedwater contains 350 ppm of total solids, a blowdown rate of 10% would be required to maintain the solids level in the boiler at 3500 ppm. The heat loss associated with 10% blowdown is roughly 3% if none of the sensible heat carried away in the blowdown water is recovered.

Nearly all of this heat can be recovered in a continuous blowdown system incorporating heat exchangers which cool the blowdown to within 20° to 30°F of incoming makeup water. The potential saving is proportional to the amount of blowdown required which further depends on the percentage and purity of makeup water.

Pre-boiler treatment of water to reduce solids can also reduce blowdown requirements. A firm specializing in industrial water treatment would provide savings. The type of equipment and treatment costs vary widely with the specific impurities involved and are best handled on an individual basis.

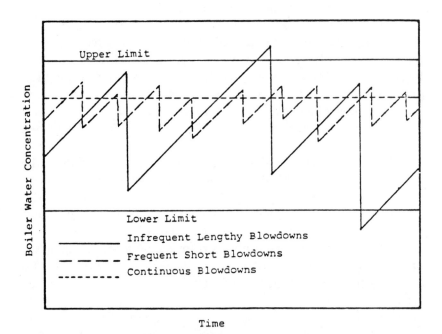

FIGURE 11-1. Variation of boiler water concentration with infrequent lengthy blowdowns, frequent short blowdowns and continuous blowdowns.

12

Effect of Water Side and Gas Side Scale Deposits

Scale provides resistance to the heat transfer between the hot flue gases and the boiler water. The heat transfer through the scale can be expressed as

$$q = k \ \frac{\Delta T}{L}$$

where L = scale thickness
k = thermal conductivity
ΔT = temperature differential

Therefore, the quantity of heat energy transferred is proportional to the thermal conductivity of the scale and inversely proportional to the scale thickness.

The thermal conductivities of several substances characteristic of boiler scale deposits are presented in Table 12-1.

TABLE 12-1. Thermal conductivities of various substances.

Material	Thermal Conductivity, Btu/Ft^2)(H)($^\circ$F)(In)
Analcite ($Na_2 O \cdot Al_2 O_3 \cdot 4SiO_2 \cdot 2H_2O$)	8.8
Calcium phosphate	25
Calcium sulfate	16
Magnesium phosphate	15
Magnetic iron oxide	20
Silicate scale, porous	0.6
Fire brick	7
Insulating brick	0.7
Boiler steel	310

WATER SIDE SCALE

Scale material in the form of dissolved solids is deposited on the water side tube surfaces during the process of water evaporation. The steam forms as bubbles on the tube surfaces. Dissolved material may precipitate from these bubbles since their surfaces contain a higher concentration of solids than do the main body of water.

The maximum allowable concentration of dissolved solids in the boiler water is determined by the operating pressure as given in Table 12-2.

The concentration of dissolved solids is generally controlled by feedwater pretreatment procedures and by blowdown. Blowdown refers to removing a portion of the boiler drum water in excess of the recommended solids concentration and replacing it with makeup water of a lower concentration. The percent of blowdown required to maintain a given dissolved solids concentration is presented in Figure 12-1.

TABLE 12-2. Recommended maximum impurity concentrations.

BOILER FEEDWATER					BOILER WATER	
Drum pressure psig	Iron ppm Fe	Copper, ppm Cu	Total hardness ppm $CaCO_3$	Silica, ppm SIO_2	Total alkalinity,[1] ppm $CaCO_3$	Specific conductance, micromhos/cm
0-300	0.100	0.050	0.300	150	700[2]	7000
301-450	0.050	0.025	0.300	90	600[2]	6000
451-600	0.030	0.020	0.200	40	500[2]	5000
601-750	0.025	0.020	0.200	30	400[2]	4000
751-900	0.020	0.015	0.100	20	300[2]	3000
901-1000	0.020	0.015	0.050	8	200[2]	2000
1001-1500	0.010	0.010	ND[4]	2	0[3]	150
1501-2000	0.010	0.010	ND[4]	1	0[3]	100

Reprinted from *Power* magazine.

Since water side scale deposits retard the transfer of heat energy to the boiler water, reduced operating efficiencies occur. The approximate energy loss as a function of scale thickness is presented in Figure 12-2 for several scale materials.

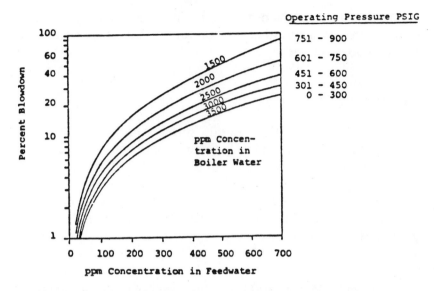

FIGURE 12-1. Percent blowdown required to maintain predetermined
boiler-water dissolved solids concentration.

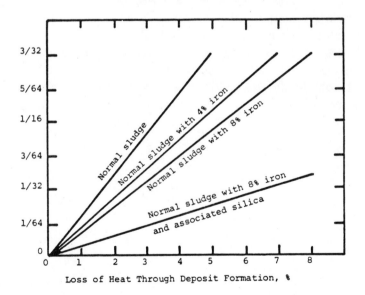

FIGURE 12-2. Energy loss from scale deposits.

Water side scale deposits also result in increased tube metal temperature often resulting in tube failure prior to a noticeable effect on boiler efficiency.

GAS SIDE SCALE

Solid deposits on the gas side surfaces of boiler tubes also result in reduced heat transfer to the boiler water. These solid deposits exist as either soot particles from the flyash or slagged ash particles.

These problems are generally related to oil- and coal-fired systems. The properties of the fuel, the characteristics of the firing system and the resulting temperature distribution in the boiler have an effect on the type and accumulation rates of the deposits.

Soot deposits can form on the furnace walls of gas-fired boilers due to excessively low excess air levels and flame impingement on the water walls. Severe soot deposits can decrease operating stack gas temperatures. Generally, a gradual increase in stack temperature is an indication of soot buildup.

Excessive gas side deposits can generate abnormal gas flow patterns in the boiler convective region and can even "plug up" a unit completely. Wall blowers are used to remove slag deposits from the furnace walls of coal-fired units. Sootblowers are used to remove flyash from convective passes of the boiler.

13

Load Management

Load management is an effective tool in minimizing fuel consumption in a plant with several boilers. Since boiler efficiency varies with load, boiler design, equipment age, fuel, and other factors, there is an advantage to distributing load to the most efficient boilers and to operating boilers at a load close to their efficiency.

In general, a peak operating efficiency will occur at a load less than 100% due to the complex interaction between the stack gas temperature, excess air flow and surface radiation losses. Units with high excess air requirements or significant radiation losses at low loads will have peak efficiency at a high load. Boilers with constant excess air levels and small radiation losses over the load range will exhibit a peak efficiency at a lower load.

Several typical efficiency curves are presented in Figure 13-1. It is usually necessary to determine the efficiency variation with load for each individual unit by measurement.

Proper load management would then require the use of the optimum operating range for each boiler in the system whenever practical and maintaining this load to minimize efficiency losses during load variations.

Plants with several boilers should be managed to achieve optimum system performance. The most efficient boilers should be loaded first to the desired operating level and so on to the least efficient boiler. Conversely, the least efficient boilers should be removed from service first.

There also is merit to shutting some boilers down in swing seasons while operating process boilers at peak efficiency as opposed to operating all boilers at reduced load and performance. Another variation in multiple boiler plants would be to schedule one boiler for a specified process, another for space heating and so on.

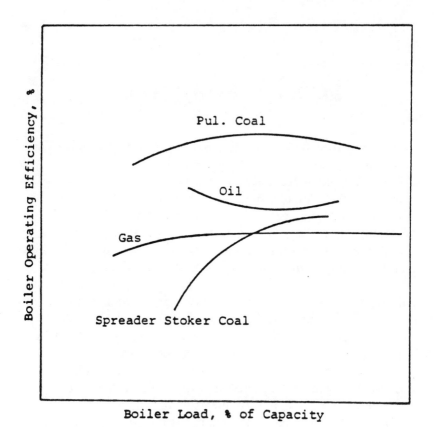

FIGURE 13-1. Typical boiler efficiency vs. load dependence.

A detailed cost analysis including additional operating personnel and maintenance must be conducted to determine the net overall cost savings in multiple boiler systems.

Proper load management also involves the use of reduced operating levels or shutdown during non-production periods. The choice between maintaining the unit at idle position, allowing the unit to be at standby at a reduced firing rate or shutting the unit down is determined by the length of the interruption, the heating and cooling characteristics of the unit, and the effects of the thermal cycles on the life of the unit.

Other considerations such as limited turndown range due to fuel firing characteristics or furnace dimensions can also dictate a preliminary shutdown rather than maintaining a low idle firing rate. Reduced operations are always justified from an efficiency standpoint if the energy saved by reduced operation is greater than the energy required to reheat the unit back up to operating conditions.

Selection of new equipment such as quick steaming units that can come to operating temperature and pressures from reduced firing levels in minutes or boilers with maximum efficiency in the range of steam flow utilized may be considered to optimize the load management potential. Several small, more efficient units that meet the specific needs of the process could show a long-term economic benefit in terms of fuel savings over fewer larger units.

A thorough engineering evaluation of the cost/benefits and potential effects on equipment life and maintenance must precede the initiation of any load management program.

FUEL CONVERSIONS OR CHANGE OF FUEL

A fuel change or fuel conversion is generally not a common tool for fuel conservation since the decision usually is controlled by other factors such as fuel availability, cost, or process requirements. However, fuel changes can have a substantial impact on the total cost of steam production. Since different fuels do characteristically exhibit modest differences in boiler efficiency, this factor should be taken into account in addition to raw fuel cost (on a million Btu basis).

These differences result primarily from dissimilar hydrogen content in the fuels with resulting differences in moisture content of the flue gas (moisture losses), variations in heat release rate in the furnace causing differences in steaming rates and dissimilar slagging and fouling characteristics.

Primary considerations in fuel selection are present and future trends in cost and availability. Other factors include:

- existing fuel storage and handling equipment
- existing combustion equipment and controls
- boiler cleanliness problems and sootblowing provisions
- air pollution emissions
- maintenance
- additional operating personnel and supervision.

A complete evaluation including preliminary testing should be made prior to a long-term commitment to an alternate fuel supply.

14

Auxiliary Equipment to Increase Boiler Efficiency: Air Preheaters and Economizers

This chapter reviews two important types of auxiliary equipment currently available to improve boiler efficiency. Most of this equipment can be retrofitted to existing units but in many cases it is economically more practical to purchase it with a new unit.

The fuel savings presented are based on information provided by manufacturers and generally represent constant load conditions at a high operating load. A more thorough evaluation must include efficiency and usage factor variations with respect to load and the potential impact on fuel savings. Other economic factors such as depreciation and discount rates are treated in Chapter 19.

AIR PREHEATERS

Principle of Operation

Air preheaters are employed to transfer heat energy from the stack gas to the incoming combustion air supply. Boiler efficiency is improved at all loads as a result of reduced stack gas temperatures. Improved combustion conditions at the burner permitting operation with lower excess air can be an additional benefit.

Air preheaters can be used economically on units with outputs as low as 25,000 pounds of steam per hour. The choice between an air preheater and an economizer for larger units is made on the basis of

- stack gas temperature
- installation and operating costs
- draft losses
- type and arrangement of the boiler
- operating steam pressure that may limit the maximum feed-water temperature and thereby the application of an economizer
- flue gas corrosive properties (i.e., fuel type) which may determine the lowest practical stack temperature
- maintenance requirements

A dual installation incorporating both an air preheater and economizer is often used on larger units with steam pressures above 400 psig.

Performance

Considering only the benefit of utilizing heat energy in the flue gas to heat the combustion air, the overall steam generator efficiency will increase approximately 2.5% for every 100°F decrease in exit gas temperature or 2% for every 100°F increase in combustion air temperature. Any accompanying improvements in burner excess air with preheated combustion air will enable additional efficiency gains. As an example, an air preheater providing combustion air of 300°F can result in a fuel savings of up to 5%.

The efficiency improvement for a given drop in flue gas temperature can be estimated using Figure 14-1. Using this value, an approximate gross annual fuel savings can then be estimated using the nomograph shown in Figure 14-2. The steam flow used in this nomograph should correspond to an average load based on 6000 hours/year operation. Subtracting fixed charges and increased operating costs will result in an approximate net annual savings.

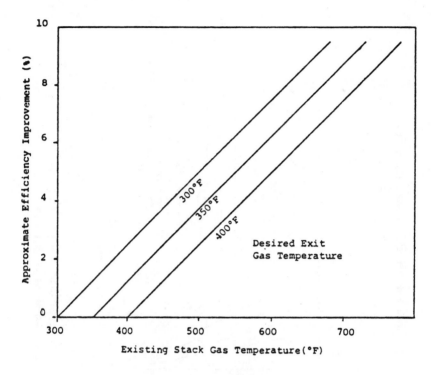

FIGURE 14-1. Approximate efficiency improvement with
decreased flue gas temperatures

Cost

The total cost of installing an air preheater will of course vary
from boiler to boiler, depending on the pressure drop that can be
tolerated, air heater design, materials and desired performance.
Installed costs can be as much as 3 to 5 times the capital costs
and are affected by

- available space
- required equipment repositioning
- duct modifications
- air fan changes if necessary to overcome increased pressure
  drops
- piping and wiring changes
- additional sootblower requirements.

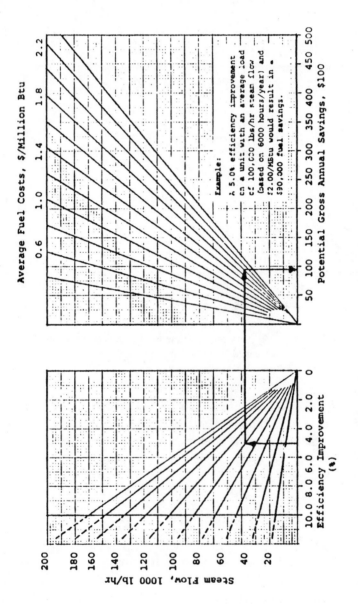

FIGURE 14-2. Potential gross savings in annual fuel costs for a given efficiency improvement based on 6000 hr/yr operation.

Significant annual costs must also be considered including fixed costs, operating and required maintenance costs. Fixed and operating costs may amount to 10-20% of the installed costs. A cold end heating surface replacement can cost another 10% of the equipment costs.

Types of Air Preheaters

Air preheaters are classified as either recuperative or regenerative. In the recuperative type, the heat energy is transferred directly from the flue gas on one side of a stationary separating surface to air on the other side. This separating surface may be either a tubular or flat plate design. Typical arrangements for the more common tube types are shown in Figure 14-3. Also included in this class are recuperative steam coil heaters that are sometimes used to preheat the combustion air before entering the main air preheater. This is occasionally necessary to reduce "cold side" corrosion by increasing the minimum tube metal temperatures exposed to the flue gas. Advantages of recuperative air preheaters include

- adaptable to a wide variety of physical arrangements
- limited cross contamination between flow streams
- no external power
- no moving parts

A disadvantage is the adverse effects of soot deposits on heat transfer efficiency. Use of the plate type is diminishing due to cleaning difficulties.

The regenerative type preheater transfers the heat energy from the flue gas to the combustion air through an intermediate heat storage medium. Steam generating units most commonly use a rotary "wheel" design shown in Figure 14-4. Advantages of this type of unit include

- adaptable to several duct arrangements
- compact and lighter than comparable recuperative units
- short flow passages reducing fouling

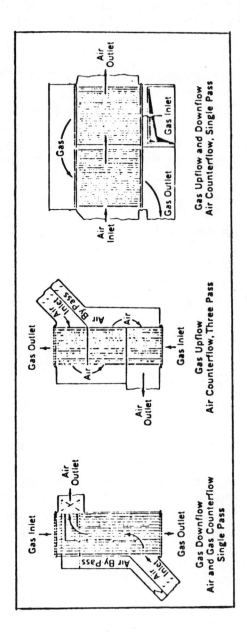

FIGURE 14-3. Typical recuperative air preheater arrangements.

- heat transfer characteristics not affected by soot deposits
- can operate at a lower exit gas temperature than recuperative units

The disadvantages are

- cross contamination around seals
- external power required

It is generally more common to use a regenerative rotary unit on retrofit installations.

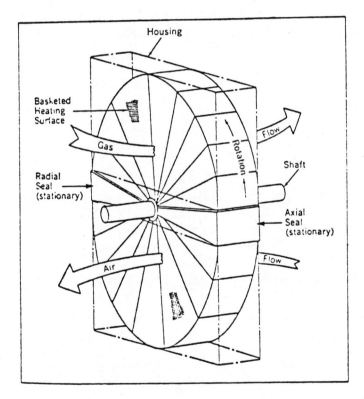

FIGURE 14-4. Regenerative air preheater.

A third type of air preheater not currently in use on steam generating units is a heat exchanger employing heat pipes as shown in Figure 14-5. Heat pipes are a relatively new technological development and offer a promising alternative to conventional heat transfer devices. This type of system may be very adaptable to industrial boilers and offers the following advantages:

- no cross contamination
- no external power or moving parts
- minimum maintenance

The primary disadvantage is a 600°F maximum operating temperature which would limit its application. Flue gas bypasses are recommended for protection from possible corrosion at low loads and overheating of the heat pipe element at high loads.

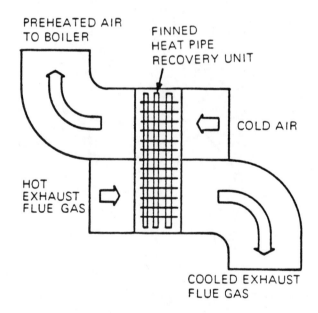

FIGURE 14-5. Heat pipe heat exchanger.

Impact on Emissions

Increased combustion air temperatures is known to have an adverse effect on emissions of oxides of nitrogen (NOx), particularly for natural gas fuel. For a range of industrial boiler designs, the NOx increased by 20 ppm to 100 ppm per 100°F increase in combustion air temperature.

This factor must be considered where existing or proposed air pollution regulations may be applicable. Improved combustion conditions in the furnace resulting in lower excess air requirements may tend to offset somewhat this increase in NOx levels since NOx usually decreases with lower excess air.

Use of preheated combustion air also appears to affect other air pollutants. Lower SO_3 emissions have been reported; however, this has not been fully substantiated. More complete combustion in the furnace may also reduce particulate and CO emission levels.

Corrosion

The lowest acceptable flue gas temperature at the exit of the air preheater is limited to prevent moisture condensation on heat transfer surfaces which would lead to sulfuric acid corrosion and expensive repairs. Factors affecting these minimum temperature limits include

- sulfur content of the fuel
- moisture content of the flue gas
- type of fuel
- ambient air temperature

Recommended minimum average cold end temperatures developed by C-E Air Preheater are presented in Figure 14-6. Note that the minimum metal surface temperatures are determined by averaging the exit gas and entering air temperature of 80°F, the exit gas temperatures corresponding to 235°F and 175°F average cold end temperature would be 390°F and 270°F respectively.

Average cold end temperature may be controlled by one or more of the following:

- Bypass a portion of inlet air around heating surfaces
- Recirculating a portion of the hot air from the air heater outlet into the air entering the heater
- Use of a recuperative steam coil in the air duct upstream of the main air preheater.

Corrosion may be further reduced by using acid-resistent metals such as cast iron and corrosion-resistant steel or by coating metal surfaces with enamel.

Ash deposits on the heat transfer surfaces also contribute to corrosion. The porous ash reduces underlying metal temperatures and allows the flue gas moisture to condense on the metal surface. The moisture that results combines with the acid in the ash to promote rapid corrosion. Soot blowers employing steam or air as cleaning agents are used to remove the ash deposits (see Chapter 15).

The life of the heat transfer surfaces in an air preheater is very dependent on the prevention of cold end corrosion and the sulfur content of the fuels used. The expected life ranges from 4 or 5 years with high sulfur fuels to 20 years on natural gas.

Maximum Combustion Air Temperatures

The temperature to which the combustion air can be preheated is limited by the entering gas temperature, the firing equipment and the type of duct, windbox and furnace construction. Maximum combustion air temperature is generally taken to be 250-350°F for stoker-fired units. Pulverized-coal firing may require preheated air up to 700°F. Retrofit applications must consider these limitations along with required insulation of combustion air ducts.

Air Preheater Size

The physical dimensions of the air preheater are determined by

- arrangement and dimensions of air and gas passages
- space available for the air preheater
- type of fuel fired
- desired final combustion air and flue gas temperature
- air and gas flow requirements
- pressure drop requirements

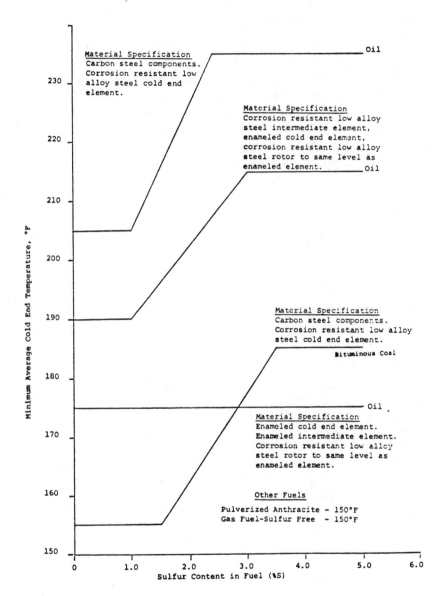

FIGURE 14-6. Cold end temperature and material selection guide.

Forced Draft and Induced Draft Fan Modifications

An evaluation of the use of air preheaters in existing units must consider the added cost of revamping forced and induced draft fans to accommodate the added resistance to air and gas flow. Static pressures will also increase due to changes in air and gas temperatures throughout the unit. Additional fan power will be required to make up for leakage around seals on a rotary type unit. Typical design leakage can be as much as 10-15% of the exit gas flow on this type of air heater while leakage on recuperative heaters can be eliminated entirely by replacing or repairing leaking elements.

A complete engineering evaluation will be required to determine the feasibility of using existing or required new equipment and the added operational costs.

Additional Desirable and Undesirable Aspects

In addition to the improvements in overall boiler efficiency, and subsequent fuel savings, air preheaters offer the following desirable features:

- Improves combustion efficiency resulting in reduced excess air requirements for complete combustion and total gas flow through the unit.
- Aids in stabilizing fuel ignition which permits greater load flexibility including improved low load operation.
- Increases furnace temperature and heat absorption rate resulting in greater steam producing capacity limited by drum steam separating capacity.
- Results in more complete burning of the fuel producing cleaner gases passing through the unit, less fouling and fewer boiler outages for cleaning.

The disadvantages of air preheaters include the following:

- Increased stoker and furnace refractory maintenance costs.
- Accumulation of deposits in the gas passages of the air heater may restrict gas flow and in extreme cases may ignite, causing serious damage (if poorly maintained).

ECONOMIZERS

Principle of Operation

An economizer is an arrangement of feedwater tubes located in the exhaust duct that absorb a portion of the heat energy that would otherwise be lost in the flue gas. This recovered energy provides additional heating of the feedwater thereby reducing the boiler firing rate necessary to generate steam, improving overall boiler efficiency. Economizers can also be employed to heat hot water for space heating or process requirements apart from the boiler water cycle.

Economizers are generally preferred over air preheaters in retrofit additions to industrial-sized units for the following reasons:

- lower initial capital costs
- no impact on NOx emissions
- lower draft losses
- minimal auxiliary power requirements (slight increases in feedwater pumping and combustion air fan requirements)

An economizer is more economical than an air preheater for small, low pressure boilers with outputs below 50,000 pounds of steam per hour. Air preheaters become competitive with economizers for larger units and the choice is made on the basis of the same factors itemized earlier in this chapter, under the Air Preheater "Principles of Operation" heading.

Performance

An increase in feedwater temperature of 10°F will result in a fuel savings of approximately 1.0%. The practical extent to which the feedwater temperature can be increased using an economizer is limited by several factors including

- flue gas temperature
- steam pressure
- economizer surface area
- corrosion characteristics of flue gas.

The efficiency improvement with an economizer is reflected in lower stack gas temperatures and can be estimated using the graph presented in Figure 14-1. The relationship between stack temperature reduction and feedwater temperature increase can be closely estimated on the basis that a three-degree decrease in flue gas temperature will raise the feedwater temperature by one degree. There is a maximum feedwater temperature depending on the operating pressures of the boiler. This constraint is discussed later, under the heading "Economizer Outlet Temperature."

The gross annual fuel savings can be estimated from the nomograph presented in Figure 14-2 using the projected efficiency improvement, average steam flow and fuel costs. Subtracting fixed charges and increased operating costs will result in an estimated net annual savings.

Costs

Equipment and installed costs of retrofitting an economizer into an existing steam generator will vary for each application. Capital costs are influenced by such factors as

- arrangement of economizer tubes
- desired performance
- tube size and materials

Installed costs will typically vary from 30 to 100% of the capital costs depending on:

- available space
- duct modifications
- required feedwater preheaters (below, under heading "Economizer Outlet Temperature"
- plumbing and pumping alterations
- air fan changes due to increased draft loss

Maintenance costs are generally limited to internal and external cleaning. Tube replacement costs can be minimized by proper selection of economizer tube materials for given flue gas conditions as outlined below.

Size

Heat transfer rates are highest when the temperature differential between the flue gas and the feedwater are greatest. Since the feedwater is at a lower temperature than the water in the convective pass tubes of the boiler, absorption of heat from the flue gas is potentially more effective in the economizer than in the convective surfaces of the boiler.

Current boiler design trends have made use of this fact by increasing economizer surface with corresponding decreases in boiler surface. The economizer surface area is generally 25 to 30% and 40 to 60% of the boiler heating surface for applications with and without air preheaters respectively. Economizer surface area is limited by increased draft losses and outlet temperature differential between the gas and feedwater.

Arrangement

Economizers can be classified as either integral or separate. Integral economizers are arranged in vertical or horizontal banks of tubes located within the boiler casing. Vertical tubes are similar to a bank of boiler tubes and are frequently connected at both ends with drums. Horizontal tubes are arranged in closely spaced, staggered rows.

Spacing is limited to a minimum of 1¾-2 inches to prevent ash accumulation from plugging the economizer and to maintain draft losses to less than 4 inches of water. Horizontal tubes have a continuous loop arrangement.

Separate economizers are located downstream of any convective region and outside the boiler casing. They are generally arranged horizontally. Retrofit economizers usually employ separate economizers.

Water flow in the economizer is normally from the bottom upward through the tubes allowing any steam that may be generated to be bled off the top. The flue gas passes either down along the tubes moving counter to the water flow or across the tubes in single or multiple passes. The counter flow arrangement is preferred as it reduces required surface area and results in lower draft losses. Baffles and bypass ducts are sometimes employed to control gas and feedwater temperatures.

Tubes

Economizer tubes are generally made of cast iron or steel and range from 1¾-2 inches O.D. Steel tubing is employed in high pressure applications and offers the following advantages

- thinner tubing
- closer spacing
- higher conductivity
- lower initial cost

Cast iron is used in low pressure applications especially where corrosion is expected. A combined steel and cast iron design is sometimes used where maximum resistance to both internal and external corrosion is desired.

Extended tube surfaces are often used to increase heat absorption. The type of extensions and spacing are determined by the fuel characteristics and provision for cleaning. Short, widely spaced fins are employed with coal and residual oil while tall, closely spaced fins are used with natural gas or other clean fuels. The advantages of this type over bare tube surfaces include

- lower initial costs
- strength against bending
- reduced tubing requirements.

Extended surface tubes are normally installed in horizontal banks in a staggered arrangement.

Economizer Outlet Temperature

The maximum economizer outlet temperature is usually maintained 35 to 75 degrees below the saturation temperature to prevent steaming, water hammer and thermal shock. In certain applications, steaming economizers are employed to produce part of the steam. They are not normally employed in industrial boilers where a high feedwater makeup is necessary.

In the case of retrofit economizers significant feedwater temperature alterations can have a detrimental effect on boiler water circulation. Industrial boilers generally use "natural" water circulation (i.e., circulation throughout the boiler is maintained by variations in water densities caused by heat absorption in various parts of the boiler circuit). Changes in feedwater temperature can result in a lower circulating head and some circuits may not maintain sufficient flow for proper cooling. This may require modifications to existing boiler heat transfer passes.

Internal Corrosion

Economizers are subject to internal corrosion from dissolved oxygen and low hydroxyl ion concentration. As the temperature of the feedwater increases, the oxygen is driven out of solution and attacks the tube interior surfaces. The feedwater is therefore preheated to 220°F by the use of deaerator heaters to remove dissolved oxygen. A pH value of between 8 and 9 is normally maintained in the feedwater by recirculating boiler water into the inlet of the economizer.

External Corrosion

The same external corrosion considerations that were discussed in the Air Preheater section apply to the application of an economizer. Since the tube metal temperature is very close to the feedwater temperature, the inlet feedwater temperature is normally maintained at 220°F or higher so that tube metal surfaces will be above the minimum cold end temperatures presented in Figure 14-6. Exit flue gas temperatures are at least 80°F above this to provide sufficient heat transfer to the feedwater.

Air Pollution Emissions

Air pollution emissions are generally not affected by the use of an economizer. CO, particulate and combustible emissions may be reduced in applications where overfiring of the boiler was necessary to maintain peak load conditions prior to incorporation of an economizer.

15

Other Types of Auxiliary Equipment

FIRETUBE TURBULATORS

Principle of Operation

Firetube turbulators are small baffles inserted into individual firetubes in the upper passes of firetube boilers to induce turbulence in the hot gas stream and thereby increase the convective heat transfer to the tube surface. These devices have also reportedly enabled installers to balance gas flow through the firetube to achieve more effective utilization of existing heat transfer surface. The corresponding efficiency improvement at a given boiler firing rate and excess air level are reflected in lower stack gas temperature and increased steam generation.

Performance

As described in the previous chapter, a reduction of $100°F$ in stack gas temperature will result in a $2\frac{1}{2}\%$ efficiency improvement. Manufacturer claims range up to 10%, which would correspond to a $400°F$ reduction in flue gas temperature. These very high improvements are probably a combined effect of reducing excess air levels, lower stack gas temperatures and tube cleaning which may be performed during installation of the turbulators.

Physical Description and Application

Turbulators may consist of angular metal strips inserted into the second or second and third flue gas passes or spiral metal blades and baffles installed at the tube inlet. A typical installation of the first type is depicted in Figure 15-1. These are generally 2 to 8 feet in length and sized to fit the inside diameter of the tubes.

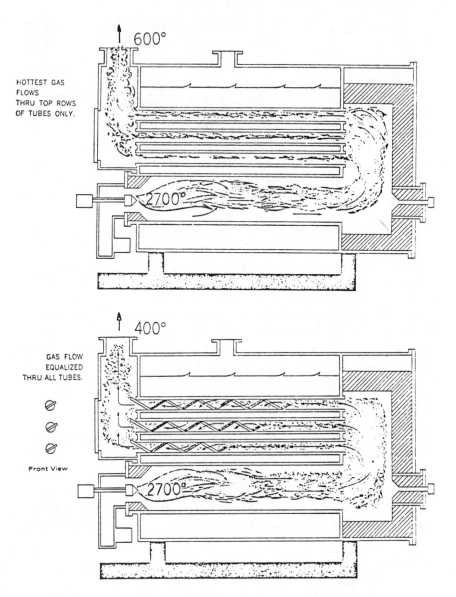

FIGURE 15-1. Two-pass Scotch Marine firetube boiler with and without turbulators.

These devices have been generally employed on firetube units with steam capacities less than 30,000 pounds per hour as substitutes for more costly economizers or air preheaters. Turbulators can be employed in gas- and oil-fired boilers with two and three pass systems but are not recommended for coal-fired units.

Turbulators increase convective heat transfer to the tube surface by increasing flue gas turbulence. By selectively altering the flow resistance through individual tubes, they can also be employed to balance gas flows through the tubes.

Allowances must be made to overcome the increased pressure drop through the system and assure sufficient air flow for complete combustion. Often this will require readjustments of burner controls following installation of the turbulators. The need for these adjustments will be determined by a post-installation efficiency check (O_2, CO, stack temperature measurements) which should be performed to verify complete combustion.

Air Pollution Emissions

Manufacturers claim that the improved combustion conditions will result in reduced combustibles in the flue gas.

OIL AND GAS
BURNERS AND SUPPLY SYSTEMS

This section describes the available auxiliary equipment that can be employed to improve burner performance for oil- and gas-fired systems.

Principle of Operation

Excess air levels have a major impact on boiler efficiency. Too much excess air means higher NOx emission levels, increased mass flow, and energy losses up the stack. The additional air results primarily from excess air at the burner required to provide complete combustion of the fuel. Current design trends have employed modified fuel/air mixing and flow patterns to minimize excess air requirements. Associated reductions in NOx and particulate emissions have also been demonstrated.

Low excess air (LEA) burners originally developed in Europe are currently being applied to industrial steam generating units in the United States. These combustion systems employ improved oil fuel atomization and controlled air fuel mixing to achieve extremely low excess air (3-5%) operating levels.

Many of these systems were initially developed to reduce air pollution levels and can therefore offer this additional benefit. Some restrictions, however, have been encountered in the application of LEA burners to industrial boilers especially on a retrofit basis.

These problems have been primarily associated with the compatibility of new flame shapes and heat release rates with existing furnace volume and design. A very sophisticated combustion control system employing automatic O_2 correction is required with an LEA burner to fully utilize its low excess air capability and prevent small excursions in excess air which might lead to high combustibles and potentially unsafe conditions.

Conventional packaged burner systems that come complete with fans and combustion controls are also available. While these systems are not as sophisticated as those described above, they can result in substantial reductions in required operating excess air levels where existing burners require high excess air due to outdated design, wear, deterioration, etc. They also have the advantage of reduced maintenance, convenient replacement and are readily adaptable to nearly all industrial boiler configurations.

Other equipment that can benefit boiler efficiency includes special oil fuel atomization and viscosity control systems. Proper pretreatment and atomization of the oil fuel are vital to reduced excess air operation.

Performance

The performance benefits using improved combustion equipment are determined by the resulting reduction in operating excess air levels.

Several manufacturers of LEA burners have claimed operating excess air levels as low as 2%. It has been questioned, however, whether these excess air levels can be maintained throughout the entire load range and over prolonged operation especially on multiple burner systems.

Combustion control systems use an average flue gas excess O_2 level measured at the boiler outlet to adjust all the burners as a single unit. Variations in air or fuel flow between burners due to normal oil tip wear, gradual dissimilarities in air damper adjustments, etc. may result in the inability to maintain extremely low excess air levels without extensive maintenance requirements. More realistically attainable levels may prove to be approximately 5% excess air.

Manufacturers of packaged burner systems guarantee an operating level of 10% excess air for both oil and gas fuels.

LEA burners generally resemble conventional burners. Modifications in fuel injection and air flow patterns to achieve optimum mixing and flame patterns permit very low excess air while maintaining complete combustion. As with conventional burners, the minimum excess air levels are indicated by excessive levels of CO or hydrocarbons on gas fuel and smoke on oil fuel.

Current design trends have emphasized the use of primary and secondary combustion zones within the flame to minimize NOx formation. The primary zone nearest the burner is maintained fuel rich with fuel and air mixing characteristics designed to provide stable flame holding. Swirl or rotation in the air stream or, in certain cases, turbulence from deflectors, provide the necessary flow pattern. Additional air required for complete combustion is provided in a secondary combustion zone surrounding the primary zone.

The combustion of oil fuels involves an additional aspect not present in the combustion of gaseous fuels. Oil fuel must be atomized prior to mixing with the combustion air. This is accomplished by dispersing the oil into a fine droplet spray that evaporates from the heat of the flame. The fineness of the atomization determines the evaporation rate of the oil and will vary with the viscosity of the oil entering the atomizer.

LEA burners develop the fuel spray and air flow patterns to achieve both complete mixing and a stable flame pattern while requiring less overall excess air than conventional burners. Fuel sprays are oriented or swirled in several fashions to direct the oil vapors into the air flow.

Various air flow patterns are employed to provide a controlled mixing pattern for complete combustion with minimum excess air requirements. This flow pattern holds the mixture in proper position in front of the atomizer and provides for vaporization and ignition of the oil droplets by locally recirculating exhaust products and by radiation from the flame.

Numerous factors must be considered in the selection of a new burner system. Flame shape and heat release rates must be compatible with furnace conditions. Use of a system that can employ existing fan and fuel handling capabilities will significantly reduce costs. Adequate turndown capabilities and firing rate response for load modulations are important.

Flame scanners and other safety equipment and instrumentation may be useful to improve reliability. A comprehensive engineering evaluation and consultation with the boiler manufacturer is recommended prior to the selection of new burner equipment.

Atomizing Systems — Efficient combustion of a liquid fuel requires rapid vaporization of the fuel which is facilitated by atomizing the fuel to the smallest possible droplet size.

Many new concepts for improved fuel oil atomization are currently available. One system employs sound waves to "sonically atomize" the fuel to a fine liquid spray. Another system uses water emulsion in the oil that flashes to steam at the burner nozzle and shatters the oil into small particles. The manufacturers of these techniques generally claim improved turndown ratios and as much as a 5% savings in fuel when compared to the conventional steam, air, mechanical or pressure atomizers currently utilized in industrial boiler burners.

When evaluating new atomizing systems it should be recognized that these will impact boiler efficiency primarily by permitting operation at lower excess air levels thereby minimizing the "dry gas losses." For retrofit applications these systems are therefore directed to existing burners which are of inadequate design to operate with low excess air.

Viscosity Control — Viscosity is a measurement of the relative ease by which a liquid will flow when subjected to a pressure differential. Viscosity is important to burner performance in the direct relationship it has with atomization. Droplet size, spray cone angle, pressure drop, and soot formation in the burner are all affected by the viscosity of the oil. Proper atomization is basic to achieving complete combustion with low excess air.

For optimum operation, all oil burners depend on an accurate control of the oil's viscosity. Common practice is to control oil temperature using a known relationship between temperature and viscosity. Recent viscometer developments have made direct viscosity control possible. Manufacturers of such instruments claim that direct viscosity control is now necessary due to the wide variations in oils currently on the market.

Air Pollution Emissions — One of the primary factors in the development of low excess air burners has been the desire to reduce NOx and particulate emission levels. Manufacturers claim that up to a 20% reduction in NOx emissions can be achieved using a LEA burner with corresponding reductions in particulate levels.

Lower NOx emission levels can be expected with low excess air operation on any industrial burner. Typically, NOx levels are reduced 20-100 ppm for every 1.0% decrease in excess O_2.

WATER SIDE WASTE HEAT RECOVERY

Principle of Operation

Heat energy lost from the steam generating unit in the blowdown water and expelled condensate can represent a significant amount of wasted energy. While not contributing directly to boiler efficiency (i.e., extraction of heat energy from the fuel is not affected), recovering this lost energy can result in very substantial fuel savings. For this reason heat recovery from blowdown and condensate is included in this book.

Blowdown is a customary procedure to remove boiler water impurities that can affect steam quality and result in tube scale deposits. As discussed below, the amount of hot drain water discarded during blowdown is dependent on boiler and make-up water quality and can be as much as 5 to 10% of total boiler steam flow. A large portion of blowdown heat energy can be reclaimed by continuous blowdown extraction. With this technique flashed steam from the blowdown water is recycled to the boiler feedwater and heat exchangers remove heat energy from the remaining blowdown water. This heat recovery is practical only with continuous blowdown. Reducing blowdown requirements by de-ionizing and other boiler water pretreatment procedures can also minimize blowdown losses.

Recovering heat energy normally lost in steam condensate can also reduce fuel consumption. Similar to blowdown heat recovery this will not affect boiler efficiency as it is customarily defined but can lead to improved overall plant or system efficiency. Various systems are available as outlined below to collect and return the condensate to the boiler. Fuel savings will be realized by recycling as much condensate as possible at boiler operating temperature and pressure. Heat losses in the condensate system are encountered with discarded condensate, pressure drops through the system in traps and receivers, flashed steam in vented systems and cooling requirements for condensate pumps. Contamination and losses will limit the amount of condensate that can be recycled.

Performance

Typical energy savings that can be generated by recovering heat in blowdown water can be estimated using Figures 15-2 and 15-3. The percentage of blowdown will be fixed by the solids concentration of the make-up water and the maximum boiler water concentration.

Use of continuous rather than intermittent blowdown alone saves treated boiler water and can result in significant energy savings as discussed in Chapter 11. Typical intermittent blowdown

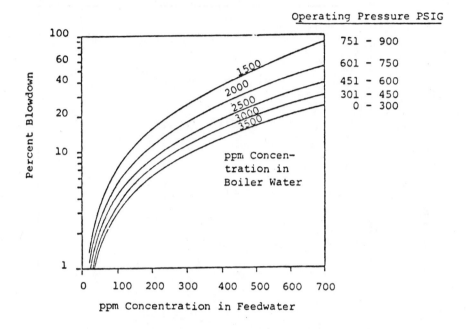

FIGURE 15-2. Percent blowdown required to maintain predetermined
boiler-water dissolved solids concentration.

patterns shown in Figure 11-1 can result in periodic shifts in total
solids concentration and require higher blowdown rates to avoid
solids carryover with the steam. Actual fuel savings in changing
over to continuous blowdown depends on the existing blowdown
patterns. The use of continuous blowdown will also permit the
recovery of blowdown heat energy described above.

Manufacturer estimates of fuel savings using semi-pressurized
and fully pressurized condensate return systems as compared to
open systems are approximately 12% to 15% respectively. Actual
savings will depend on boiler operating pressure, flash tank pressure and the eventual use of the flashed steam. Preheating boiler
feedwater with waste heat can amount to a 1% savings for every
10°F increase in feedwater temperature.

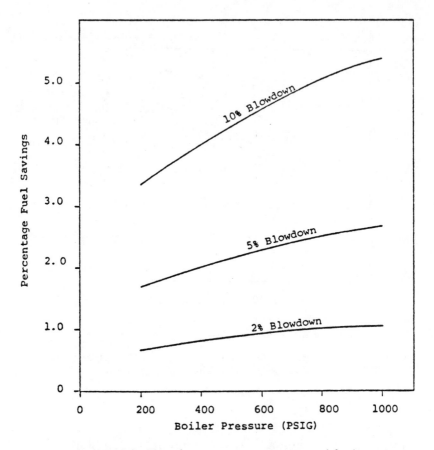

FIGURE 15-3. Waste heat energy recovery potential using
continuous blowdown system, flash steam heat recovery
and heat exchangers on expelled blowdown water.

Blowdown — Table 12-2 in Chapter 12 presents the recommended maximum concentration of impurities that can be tolerated in the boiler above which priming, foaming, and carryover occur, resulting in low quality steam. Excessive concentrations of dissolved and suspended solids also lead to internal scale formation, reduced heat transfer and, ultimately, tube failures. Boilers are therefore equipped with blowdown systems that remove boiler water with high impurity concentrations and replace it with clean make-up water. Make-up water de-ionizers can also be employed to remove dissolved minerals, thus reducing blowdown requirements.

Intermittent blowdown may result in wasted hot water and wide fluctuations in impurity concentrations. Improved operation is attained using continuous blowdown systems operating either manually or automatically that remove a small quantity of boiler water to maintain relatively constant dissolved solids concentration. Significant savings in hot water can be attained using continuous blowdown.

In addition, continuous blowdown allows the use of heat recovery devices to preheat the incoming make-up water with waste heat from the blowdown water. It is common practice to recover as much of the heat in the expelled hot water as possible. Low pressure boilers with low blowdown rates usually employ heat exchangers to preheat the incoming make-up water. Higher capacity units operating at high pressures use a flash system in which a portion of the waste water is flashed to high purity steam at low pressures, condensed and recycled to the feedwater system. The remaining waste water having now an even higher concentration of impurities is passed through heat exchangers to preheat make-up water and discarded.

Condensate Return Systems — Condensate recovered from process steam at boiler operating pressure and temperature represents a valuable quantity of clean water and usable energy. The amounts of condensate that can be recovered are dependent on the degree of contamination in the condensed water. In certain cases separators or filters can be employed to remove contaminants such as oil or rust material so that the water is suitable for use in the boiler. Water containing a high degree of contaminants is generally discarded.

The recovery of condensate is also limited by the type of system utilized. The technique employed to recycle the condensate to the boiler can be divided into five categories:

- Atmospheric (open) system
- Semi-pressurized (semi-closed) system

- Fully pressurized (closed) system
- Vacuum type
- Deaerators

A study must be conducted in any retrofit application to assure that the system chosen is compatible with existing boilers and process requirements.

The atmospheric system employs a vented tank downstream of a steam trap or drain in which the hot condensate is received. This is done to reduce condensate temperature to 212°F to allow pumping without cavitation occurring. A portion of the condensate will flash to steam and vent to the atmosphere representing a loss of both energy and water. Treated make-up water is added as required to meet feedwater flow requirements.

This is the most common type of return system with the least cost. It is generally compatible with all units especially when large amounts of condensate are returned. Losses may be reduced by proper sizing and application of steam traps and drains.

The semi-pressurized system employs a condensate receiver that is at an intermediate pressure at least 50 psi below the lowest anticipated steam pressure in the process system. The intermediate pressure permits use of conventional condensate (trap) return systems, while retaining a significantly higher condensate temperature. The receiver pressure is controlled by steam passing through a pressure-reducing valve from the main steam header to the receiver.

This steam, along with the steam flashed from the condensate, can be employed to heat make-up water or supply low-pressure process steam. A steam relief valve is provided with a set point 10 to 20 psi above the pressure reducing valve setting and prevents abnormal pressure surges. An air vent is provided on the receiving vessel to allow noncondensable gases to escape.

The condensate is cooled by the make-up water to a temperature at least 20°F below the saturation temperature to prevent vapor lock in the boiler feedwater pump. This system has the advantage of reduced water and energy losses. Pressure variations must be controlled to a minimum in the receiver in order to prevent boiler feed pump cavitation.

Relatively stable pressure can be maintained by the steam pressure-reducing valve and the steam relief valve combination— even in the face of irregular condensate flow and systems steam trap malfunctions.

An attempt is made to return all the condensate to the boiler in the fully pressurized system. The entire system is operated at the steam process pressure with flow losses overcome by specially designed water pumps.

Operation of a fully pressurized system is restricted to units with nearly constant steam loads as sudden heat demands will cause pressure fluctuations resulting in the flashing of condensate and restricted flow through the system. Vents operating automatically are provided to allow trapped air to be released. These severe operating restrictions often limit its application.

The vacuum type system is normally only used on low-pressure heating systems where large pressure drops exist due to long condensate lines or steam traps.

Deaerators preheat feedwater and remove non-condensible and corrosive gases using heat and agitation. Heat is supplied by returned condensate or low-pressure steam while agitation is provided by sprays or trays. Deaerators normally operate at approximately 15 psig. The small amount (1 to 2%) of flashed steam not condensed with the feedwater is vented with the non-condensible gases and later condensed in a vent condenser.

WALL AND SOOT BLOWERS

Principle of Operation

Wall and soot blowers are employed to remove deposits from heat transfer surfaces in the boiler. Removal of slag deposits from the furnace walls of a coal-fired unit is done using wall blowers. Soot blowers are employed to remove fly ash and soot deposits from the convective passes of the boiler. If allowed to accumulate, these deposits would retard heat transfer and may eventually result in clogging the boiler.

These problems are related to oil- and coal-fired systems. The properties of the fuel, the characteristics of the firing system, and the resulting temperature distribution in the boiler have an effect on the type and accumulation rates of the deposits.

Analyses of these factors are required prior to selection and installation of a soot-blowing system. Retrofit applications are often required with changes in fuel supply or when existing equipment is deemed inadequate due to excessive stack gas temperatures or tube failures.

Performance

Manufacturers claim that proper soot blowing can increase unit efficiency by up to 1%.

Physical Description and Application

Soot blowers can be either fixed or retractable and use either air or steam as the blowing media. Fixed position systems are employed in low temperature regions whereas retractable systems, known as lances, are generally used in high temperature regions.

Special nozzles and arrangements are employed including rotating lance blowers as illustrated in Figure 15-4. Wall blowers are generally short retractable lances that penetrate the boiler furnace wall and rotate 360 degrees.

The availability of steam and compressed air and the cost of installation and maintenance of the supply system will determine the selection of the cleaning medium. The choice of the cleaning medium will also depend to a degree on the characteristics of the deposits.

Slags at the cold end of their "plastic" temperature range (liquid and solid mixture) can be chilled (solidified) and removed by air or steam blowers. Deposits that will remain plastic over a large range in temperature (greater than 50°F) may require a water spray to solidify the deposits for easier removal.

Air Pollution Emissions

Particulate emissions may temporarily increase when soot blowers are operated.

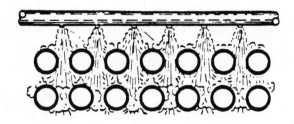

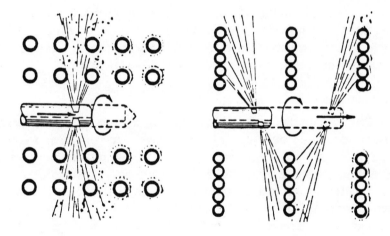

FIGURE 15-4. **Soot blower arrangements.**
Reprinted from *Power* magazine.

INSULATION

Principle of Operation

Insulation is any material that is employed to restrict the transfer of heat energy. It can generally be categorized as either mass or reflective type depending on whether it is aimed at reducing conductive or radiative heat transmission, respectively.

The overall thermal resistance of the mass type insulation depends on both its composition and physical structure. Materials are effective that have a low thermal conductivity but more importantly contain numerous small voids filled with air or gas. These voids provide the primary resistance to conductive heat transfer. The effectiveness of the insulation is also determined by the thickness or quantity of material used.

Reflective type insulation, generally consisting of smooth metal surfaces, reduce the loss of heat energy by radiation. This type of insulation is often used as a covering for mass type insulation and can also be employed in layers separated by air pockets to replace mass insulation. This application offers several advantages, however, consideration must be given to heat loss by convection and conduction through and between the metal surfaces.

Insulation provides other benefits besides reduced heat losses, including controlled surface temperatures for comfort and safety, structural strength, reduced noise, and fire protection.

Performance

Properly applied insulation can result in large savings in energy losses depending on type, thickness (mass type only), and condition of the existing insulation. Figure 15-5 presents the approximate total rate of heat energy loss by radiation and convection from a bare flat surface based on the temperature difference between the surface and still air.

Bare surface temperature in steam generating units ranges from saturation temperature on exposed tube surfaces to air and gas temperatures on duct surfaces. Approximate radiation losses from furnace walls as developed by the ABMA were presented in Chapter 3, Figures 3-3 and 3-4.

Radiation losses tend to increase with decreasing load and can be as high as 7% for small units or larger units operating at reduced loads. More precise determination of energy losses can be made based upon existing insulation operating temperature, air flow over the surface, surrounding conditions and surface orientation (horizontal or vertical). An engineering analysis of existing conditions and potential energy savings should be conducted prior to application of insulating material.

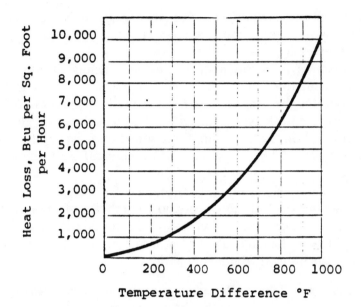

FIGURE 15-5. **Heat energy loss from bare surface.**
Reprinted from *Power* magazine.

Costs

Procedures have been developed by the Thermal Insulation Manufacturers Association to determine the optimum insulation thickness for various applications based on

- fuel costs
- operating temperatures
- insulation type
- depreciation period of plant and insulation
- capital investment

Physical Description and Application

Mass Type — The selection of an insulating material is based on several factors other than composition. As mentioned, the numerous voids contained in the material structure provide the primary resistance to heat transmission by conduction. Mass type insulation must be protected from damage by moisture absorption or excessive compression as these will reduce the insulatory value of the material.

Common insulating materials include calcium silicate, cellular glass and diatomaceous silica. Insulation made of 85% magnesia and asbestos that was once extensively used in high pressure steam generators is no longer made and has been replaced by glass fiber materials. Temperature limitations sometimes require the use of insulating materials for effective insulation of high temperature surfaces.

Stiffeners, metal mesh, or protective jackets are employed to increase the strength and dimensional stability of the insulation. Protection against fire is an important consideration both with regard to the flammability of the insulating material and protection of the insulant from external fire.

Reflective Type — Reflective type insulation generally consists of smooth metal surfaces that have a low emissivity and retard the loss of heat energy by radiation to cooler surroundings. Table 15-1 presents the emissivity of various materials that represent the ratio of the metal's radiating power to the maximum radiation rate at that temperature. In comparison, ordinary paint or mass type insulation covering have emissivities of approximately 0.85 to 0.95.

TABLE 15-1. Emissivity of metallic surfaces.

	Polished Metal Surfaces		Oxidized Surfaces
	Temperatures		Temperatures
Metal	70°F	1000°F	Below 1500°F
Aluminum	0.05	0.075	0.10 - 0.20
Copper	0.04	0.08	0.55 - 0.75
Gold	0.03	0.05	
Iron, cast or wrought	0.20	0.25	0.60 - 0.90
Monel metal	0.07	0.10	0.40 - 0.50
Nickel	0.06	0.10	0.40 - 0.60
Platinum	0.036	0.10	
Silver	0.20	0.25	
Steel	0.20	0.25	0.60 - 0.90
Tin	0.08	—	

Use of a single layer of reflective insulation does not impede the transfer of energy by convection or conduction; however, multiple layers separated by air pockets can be employed to form an insulating barrier that is more effective than mass type insulation in certain applications. Convection losses in the air separating the layers can be reduced by seals and transverse metal barriers that restrict air movement due to buoyancy forces. Conduction losses at contact points must also be considered.

Reflective type insulation offers the following advantages over mass type insulation

- lightweight
- resists moisture damage
- resists fire damage
- easier to decontaminate
- adaptable to several configurations

The greater susceptibility to corrosive damage and heat energy losses by convection and conduction are the primary disadvantages of reflective type insulation.

Table 15-2 summarizes the operating principles and potential for improvement in efficiencies for auxiliary boiler equipment.

TABLE 15-2. Summary — Boiler efficiency improvement equipment.

Device	Principle of Operation	Efficiency Improvement Potential	Special Considerations
Air Preheaters	Transfer energy from stack gases to incoming combustion air.	2.5% for each 100°F decrease in stack gas temperature	• Results in improved combustion condition • Minimum flue gas temperatures limited by corrosion characteristics of the flue gas • Application limited by space, duct orientation and maximum combustion air temperatures
Economizers	Transfer energy from stack gases to incoming feedwater	2.5% increase for each 100°F decrease in stack gas temperature (1% increase for each 10°F increase in feedwater temperature	• Minimum flue gas temperature limited by corrosion characteristics of the flue gas • Application limited on low pressure boilers • Generally preferred over air preheaters for small (50,000 lbs/hr) units
Firetube Turbulators	Increases turbulence in the secondary passes of firetube units thereby increasing efficiency	2.5% increase for each 100°F decrease in stack gas temperature	• Limited to gas- and oil-fired units • Properly deployed, they can balance gas flows through the tubes • Increases pressure drop in the system
Combustion Control Systems	Regulate the quantity of fuel and air flow	0.25% increase for each 1% decrease in excess O_2 depending on the stack gas temperature	• Vary in complexity from the simplest jackshaft system to cross limited oxygen correction system • Can operate either pneumatically or electrically • Retrofit applications must be compatible with existing burner hardware

Instrumentation	Provide operational data		• Provide records so that efficiency comparisons can be made
Oil and Gas Burners	Promote flame conditions that result in complete combustion at lower excess air levels.	0.25% increase for each 1% decrease in excess O_2 depending on the stack gas temperature	• Operation with the most elaborate low excess air burners require the use of advanced combustion control systems • Flame shape and heat release rate must be compatible with furnace characteristics • Flame scanners increase reliability • Advanced atomized systems are available • Close control of oil viscosity improves atomization
Insulation	Reduce external heat transfer	Dependent on surface temperature	• Mass type insulation has low thermal conductivity and release heat loss by conduction • Reflective insulation has smooth, metallic surfaces that reduce heat loss by radiation • Insulation provides several other advantages including structural strength, reduced noise and fire protection
Sootblowers	Remove boiler tube deposits that retard heat transfer	Dependent on the gas temperature	• Can use steam or air as the blowing media • Fixed position systems are used in low temperature regions whereas retractable lances are employed in high temperature areas • The choice of the cleaning media will depend on the characteristics of the deposits

(continued)–

TABLE 15-2. Summary — Boiler efficiency improvement equipment. *(continued)*

Device	Principle of Operation	Efficiency Improvement Potential	Special Considerations
Blowdown Systems	Transfer energy from expelled blowdown, liquids to incoming feedwater	1-3% dependent on blowdown quantities and operating pressures	• Quantity of expelled blowdown water is dependent on the boiler and makeup water quality • Continuous blowdown operation not only decreases expelled liquids but also allows the incorporation of heat recovery equipment
Condensate Return Systems	Reduce hot water requirements by recovering condensate	12-15%	• Quantity of condensate returned dependent on process and contamination • Several systems available range from atmospheric (open) to fully pressurized (closed) systems

(end)

16

Combustion Control Systems and Instrumentation

Coauthor: M.J. Slevin, President
Energy Technology and Control Corp.

Combustion controls regulate the quantity of fuel and air flows in a steam generating unit. The primary objectives of these controls are to

- Provide adequate heat input to meet steam demands automatically
- Protect personnel and equipment
- Minimize pollution
- Minimize fuel usage

The first of these objectives is met by sufficient fuel input while the latter three are dependent on maintaining the proper air flow with respect to fuel (i.e., fuel/air ratio).

An optimum range of excess air exists for each combination of fuel, firing mode and furnace condition. Too little air flow results in combustible emissions which are hazardous, polluting and can contain considerable amounts of unused energy. The minimum safe excess air level for conventional burners is generally taken to be 10% for gas fuel and 15% for oil fuel; however, specific applications may require higher levels.

Too much excess air results in higher NOx emission levels and increased mass flow and energy losses up the stack. Heat transfer within the unit is also reduced with high excess air levels culminating in higher stack gas temperatures, wasted fuel and further performance losses.

All boilers have some type of combustion control arrangement from simple manual control to a highly sophisticated computerized set up. The choice of the type of control system is made on the basis of

- steam generating unit capacity and turndown required
- steam demands and expected fluctuations in steam flow
- expected performance levels that will require more sophisticated systems for higher operating efficiency
- costs
- pollution regulations that may require low excess air operation to minimize NOx emissions
- adequate safety interlocks.

It has been shown with the utilization of microprocessor technology to be economical to equip units as small as 10,000 pounds of steam per hour with fuel saving control systems. Caution, however, must be exercised to avoid selecting a control specification that may be operationally impractical.

Instrumentation, in contrast to combustion controls, are passive elements that present the boiler conditions as they exist but are unble to detect deviations from desired operation and make corrective action without control systems or operator involvement. Without this information, it is impossible to determine whether the equipment is operating to the best advantage or that corrective action is necessary to restore performance.

Use of a continuous oxygen analyzer to determine excess air levels is highly desirable with regard to pollution control and energy conservation. They may be used either to provide information to the operator on control system operation or as an integral part of the combustion control system. Totally automatic control is achieved with advanced forms of oxygen trim control, preferably of the true adaptive type.

Performance

The increase in boiler efficiency using a retrofit combustion control system will be dependent on the type and condition of the existing equipment. Table 16-1 presents the sources of errors and the expected excess air operating levels as per manufacturers for each of the six control systems described later in this chapter. With the simplified positioning control (jackshaft) system used as the basis for comparison, installation of a sophisticated control system that adjusts air flow in relation to fuel flow with excess O_2 correction is shown to reduce air levels on gas fuel down to 10 percent excess air or less, dependant on type and condition of burner.

The associated improvement in efficiency can be determined for gas, oil and coal fuels using Figure 16-1. The reduction in excess air described above will result in a 3 to 4% benefit in fuel savings for a unit with 600°F exit gas temperature. These values will be augmented by the reductions in stack gas temperature that accompany the lower excess air flows. For the example given above, an additional 2% reduction in fuel savings will be realized. Units equipped with economizers and/or air preheaters providing lower stack gas temperatures will have less of a savings.

Costs

In the past, a fully computerized system employing direct digital control may have had an installation cost of $100,000 per boiler. However, adaptive oxygen trim controls can be installed today for as little as $11,000 per boiler.

Oxygen analyzers employing zirconium or ceramic oxide cost approximately $3000. These units have demonstrated very low maintenance requirements.

Combustion Control Systems

The six basic combustion systems employing a single fuel are described below.

Fixed Positioning — A simplified fixed positioning control (jackshaft) system has been extensively applied to industrial boilers based on minimum control system costs. A single actuator moves both the fuel and air control devices through a mechanical linkage to a preset position in response to a change in the steam pressure.

TABLE 16-1. Differences in performance of control systems.

Factors affecting accuracy of fuel/air ratio control	Fixed positioning		Positioning with operator trim		Pressure ratio		Fuel and Air metering		Cross-limited metering		Oxygen correction		Effective corrective action
	Min.	Max.	Min.	Max.	Min.	Max.	Min.	Max.	Min.	Max.	Min.	Max.	
Fuel valve and positioning linkage	1	4	1	4	–	–	–	–	–	–	–	–	Fuel metering
Upstream fuel pressure	1	3	1	3	–	–	–	–	–	–	–	–	Fuel metering
Air damper and linkage	2	6	2	6	–	–	–	–	–	–	–	–	Air metering
Fan performance	2	5	2	5	–	–	–	–	–	–	–	–	Air metering
Alignment of burner characteristic with air differential	–	–	–	–	3	10	–	–	–	–	–	–	Fuel and air metering
Burner wear and carbonization	4	10	4	10	4	10	–	–	–	–	–	–	Fuel metering
Fuel Btu, specific gravity, and temperature variation	2	4	2	4	2	4	2	4	2	4	–	–	O$_2$ correction
Temperature, pressure and humidity errors in air metering	7	14	7	14	4	9	4	9	4	9	–	–	O$_2$ correction
Air/fuel control response lag	–	–	–	–	–	–	1	2	–	–	–	–	O$_2$ correction
Unaccounted for errors	½	1	½	1	½	1	½	1	½	1	–	–	O$_2$ correction
Control system accuracy	½	1	½	1	2	5	½	2	½	2	½	2	O$_2$ correction
Potential cumulative error	20	48	20	48	15-½	39	8	18	7	16	½	2	

Probable cumulative error or additional excess air	34	34	27	13	11	1
Add for complete combustion (gas fuel)	43	43	36	22	20	9
Probable excess air required with operator trim from combustion guide (gas fuel)	43	37	31	17	15	10
Add for complete combustion (oil fuel)	53	53	40	32	30	19
Probable excess air required with operator trim from combustion guide (oil fuel)	53	47	41	27	25	20
Type of combustion guide	—	Steam flow/ airflow or oxygen analysis	Steam flow/ airflow or oxygen analysis		Oxygen analysis	—

(end)

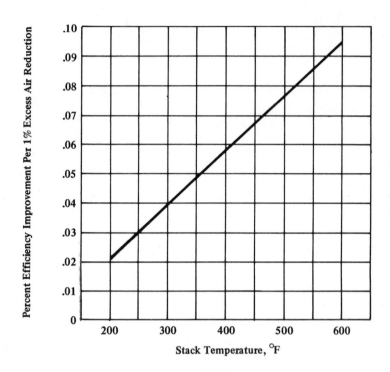

FIGURE 16-1. Curve showing percent efficiency improvement per every one percent reduction in excess air. Valid for estimating efficiency improvements on typical natural gas, #2 through #6 oils and coal fuels.

The system is calibrated by adjusting the linkage for proper fan damper travel and the profile cam on the fuel valve to produce the desired fuel/air ratio over the load range. A manual control override of the main actuator is usually provided.

Since the system merely positions fuel valve and air damper openings, it cannot compensate for changes in fuel or air density, fuel supply pressure or heating valve. Wear in the fuel valve orifice, burner tips and fan dampers will also alter fuel/air ratios. This type of control system results in considerable variations of excess air depending on the particular conditions existing at a given time.

As a result, nominally high excess air levels must be preset to avoid going below the minimum excess air requirements. In actual operation, excess air will vary over a considerable range as indicated in Table 16-1.

Parallel Positioning with Operator Trim — A lower margin of operating excess air can be achieved by substituting a pneumatic or electronic positioning system which reduces the variations caused by mechanical linkages.

Separate actuators for the fuel valve and fan damper acting in parallel from a single steam pressure controller can be added. In addition, individual manual bias control of the fuel or air input can be used to adjust the fuel/air ratio. This system requires a combustion guide in the form of a steam flow/air flow or oxygen analyzer to assist the operator in positioning the excess air flow.

The parallel positioning control is the most widely used combustion control system for units below 100,000 pounds per hour. By adjusting the fuel/air ratio, compensations for variations in fuel characteristics, combustion conditions or control system equipment can be made.

Pressure Ratio — Use of the fuel pressure at the burner and the windbox to furnace pressure differential as indications of fuel and air flows are the most elementary of the metering type systems. A basic parallel pneumatic or electronic system is used with the windbox to furnace/burner pressure ratio employed to trim the fuel and air flows. Manual bias control to change the ratio of pressures and excess air flows is normally provided incorporating a combustion guide.

This system eliminates inaccuracies due to fuel pressure and fan discharge variations but it requires that the pressure-flow relationships for the fuel and air have similar characteristics. Alignment problems often exist causing inaccuracies. Variations in fuel heating value and fuel or air densities cannot be reconciled.

Fuel and Air Metering — If the control system is further refined to include actual metering of fuel and air, additional error sources can be eliminated, further reducing the excess air levels. Several types of flow metering devices are available; however, there are some limitations as to the types of fuels that can be accurately metered.

Several combinations of components can be used in a metering control system. Feedback loops are employed to allow the fuel and air to self correct flows to meet system demands. Additional air is normally required to allow for the differences in speed of response between the fuel and air flow loops.

Cross-Limited Metering — An additional refinement to the metered system is the cross-limited metering system that limits the change in fuel flow through control logic to available air flow at all times. The quantity of air flow is also tied to the existing fuel flow and must be equal to or greater than fuel flow. Required excess air levels are thereby slightly reduced.

A common form of the cross-limited metering system is a pneumatic or electric parallel system using the steam pressure as the master controller. A manual override of the fuel/air ratio is provided to trim excess air levels.

Cross-Limited Metering with O_2 Correction — The effects on excess air of variations in fuel heating value and combustion air conditions can be eliminated using a continuous monitored flue gas O_2 level to trim the fuel/air ratio. The unit can thereby be operated at a set O_2 level resulting in minimum fuel consumption.

Combustion Control Equipment

Control systems hardware and instrumentation are either pneumatic or electronic. Considerations that must be included in the selection of equipment are

- capital costs—pneumatic systems cost 20% less than electric systems, especially final control elements
- installation costs—electric generally less expensive
- instrument air costs on pneumatic systems
- compatibility
- maintenance factors
- electronic controls are proven to be the most accurate.

Electronic components may be either analog or direct digital systems. Analog components are springs, diaphrams, cams, etc. that simulate mathematical functions used to determine the response to an incoming signal. Direct digital systems employ a calculator or microprocessor to determine the desired response and position actuators. The same considerations mentioned above with respect to costs, compatibility and maintenance would again apply to the selection of a system.

Oxygen Analyzers

Oxygen analyzers are the accepted means of determining excess air levels. Recent developments have improved the old style sampling systems with analyzers using zirconium oxide sensors so that no pretreatment of the gas sample is required. Sensors can be located directly in the flue gas stream or on the flue gas ducts connected directly to sensing probes. These systems have increased sensitivity especially in the lower excess O_2 levels and significantly reduced sample cleanliness and maintenance requirements.

Instrumentation

Table 16-2 presents the minimum instrumentation requirements to allow safe control of the boiler and provide a sufficient record of operation. Many additions to this list are possible and are subject to costs, legal requirements and personal preference of the plant operators.

Steam pressure, temperature, and flow, along with air and fuel flow, provide information on the operation of the control system in maintaining desired boiler operation at various loads. Necessary information of the boiler water supply is provided by drum level and feedwater flow indicators. Draft gauges can indicate plugging of the boiler, economizer, or air heater passages with fly ash. Gas and air temperature measurements also indicate the need for sootblower operation. Annunciators are required to warn the operator of hazardous operating conditions.

Air Pollution Emissions

Use of more sophisticated combustion controls to reduce excess air requirements will result in a reduction in NOx emission levels. Most NOx produced reacts ultimately with moisture in the atmosphere and forms nitric acid, and hence will contribute to the formation of acid rain.

TABLE 16-2. Instrument recommendations.

	Indicator	Recorder	Integrator
Boiler outlet pressure	R	R	
Superheater outlet steam temp		R	
Steam flow		R	X
Total fuel flow		R	
Total air flow		R	
Individual fuel flow		X	X
O_2 analyzer		X	
Combustibles analyzer		X	
Drum level	R	R	
Feedwater flow		X	X
Draft gauges	R		
Air and gas temp	R	X	
Feedwater temp	R		
Annunciator	R		

R Required
X Optional

Oxygen trim control data that is convenient to record will include:

- percent efficiency
- percent oxygen
- percent fire-rate
- flue gas temperature
- percent carbon dioxide
- percent oxygen error

17

Boiler O₂ Trim Controls

M.J. Slevin, President
Energy Technology and Control Corp.

For some years the market has seen a proliferation of Oxygen Trim Controls, and many have been fitted to a wide assortment of packaged boilers. Some have produced satisfactory performance and savings, but a significant number have failed to achieve anything like the salesmen's promises. In a number of cases they have totally failed to control the boiler, and have even been removed.

Many users are not sure how effective their trim controls really are. In most cases this is due to lack of information as to what the control is doing. A common criticism is the lack of adequate and precise data output from the control systems. Considerable harm has been done to the overall reputation of trim controls due in part to inadequate design and due to misrepresentation and poor application where conventional trim controls have been fitted to boilers on which they had no chance of operating correctly.

To understand why such failures have occurred, we must first understand the process that is being controlled and then determine the unique problems that manifest themselves when control is applied to the combustion process. The basis of all boilers is to transfer heat from the burning of fuel and air (combustion) into water or other liquids to produce hot water, steam, or other hot liquids.

The combustion process itself is one that is usually inefficient because of poor control of the mixture of fuel and air, and in general is adjusted for excessive air-rich combustion because of safety and the basic instability of the fuel/air ratio control mechanism. Moderate and random variations in the combustion process also occur because changes in the basic ingredients and ambient conditions are taking place in the following area: Air Temperature, Humidity, Barometric Pressure, Wind Direction, Wind Velocity, Fuel Pressure, Fuel Temperature, Fuel Calorific Value, and BTU Value.

To illustrate this, for example, natural gas will fluctuate by plus or minus 7% in terms of its calorific value, or a change of just 10°F in the temperature of a fuel oil will significantly change the BTU output of the burner. These numerous random variables make control of the boiler to fine limits very difficult. Oxygen trim controls applied to boilers have additional problems due to fundamental characteristics of the control and boiler configuration (see Figure 17-1).

Figure 17-1 shows an example of a typical four-pass, fire-tube boiler with a conventional oxygen trim control. The point of control is at the intake to the burner where the trim control will make changes to the air flow with respect to fuel flow (i.e., adjust the fuel/air ratio) in an attempt to achieve a predetermined level of excess oxygen present in the exhaust flue gas. The function is further complicated because different amounts of excess oxygen are required at different fire-rates if close to optimum performance is to be achieved.

Low fire control is particularly difficult for most conventional trim controls due to the non-linear characteristics of the air damper profile. To combat this, most trim controls have a low load cut-off which in practice results in no trim correction operation below about 25% boiler load, which for many boilers is the most critical working area, particularly during the summer months. This is also the most inefficient working range where precise trim control would be most beneficial.

The detection and measurement of excess oxygen for economic and reliability reasons is made in the stack using various types of sensors, by far the most accurate and reliable device being the

FIGURE 17-1. Where the total system lag-time occurs.

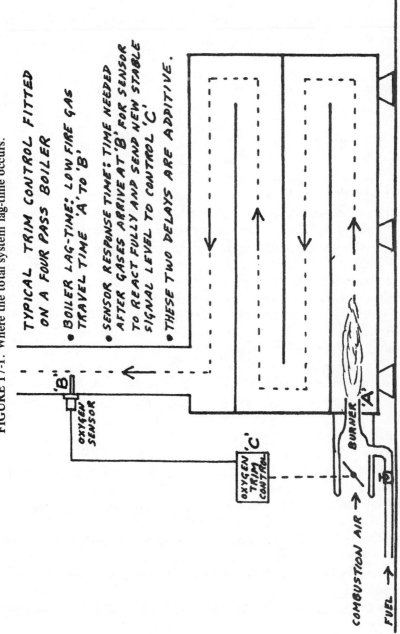

TYPICAL TRIM CONTROL FITTED ON A FOUR PASS BOILER

• BOILER LAG-TIME: LOW FIRE GAS TRAVEL TIME 'A' TO 'B'

• SENSOR RESPONSE TIME: TIME NEEDED AFTER GASES ARRIVE AT 'B' FOR SENSOR TO REACT FULLY AND SEND NEW STABLE SIGNAL LEVEL TO CONTROL 'C'

• THESE TWO DELAYS ARE ADDITIVE.

'B'

OXYGEN SENSOR

'C'

OXYGEN TRIM CONTROL

BURNER

'A'

COMBUSTION AIR →

FUEL →

zirconium oxide cell located inside the stack within the flue gas stream.

As combustion occurs at the burner the hot combusted gases travel through the passes of the boiler, transferring heat into the liquid contained within the body of the boiler, and finally the progressively cooling gases exit through the breach and pass up into the stack where they pass the oxygen sensor. The time taken for this transit of gas will vary considerably due to the fire-rate, and also due to the size, type, and design of the boiler. Typically it can be 10 to 60 seconds, and is sometimes referred to as the boiler lag-time.

Gases passing the oxygen sensor in the stack will permit the sensor to detect the change in excess oxygen, but a second time delay will occur as the oxygen sensor requires time to respond to the change in oxygen level which can be 5 to 30 seconds dependent on type, design, and capacity of the sensor.

Now we are faced with two additive delays that are in the control loop. These two delays combined could cause times from 15 seconds to well over a minute following a change at the burner due to fire-rate movement, or any other factor, before the oxygen sensor will provide an output change to the control and then a trim correction function to the fuel/air ratio to correct the oxygen to the required setpoint value.

In extreme cases of frequent fire-rate movement, the trim control never succeeds in catching up with the boiler movement and can in some situations literally make boiler performance worse than it would be without the trim control to control the operation.

Now that we have defined the lag-time problem and how it is created, what is the effect in practice? Modulating boilers are designed to raise and lower the fire-rate automatically in order to maintain a fairly constant temperature of water or pressure of steam and to compensate for changes in the output load.

There is a misconception with many boiler users that because their load is almost constant, they believe the fire-rate remains constant when in fact it moves up and down many times an hour, even on a constant load, although the amount by which it changes may be as little as 5 or 10%. Statistics obtained from on-site surveys of hundreds of modulating packaged boilers show the average number of fire-rate control movements to be around 30 per hour, but in unstable load applications we have observed the highest to be 190 movements per hour.

Each time the fire-rate moves, the conventional oxygen trim system will require a lag-time before it can make a trim correction, and in most cases it requires a second lag-time before it is likely to trim the oxygen exactly due to undershoot or overshoot on the first trim action. What this means in practice is that for an average boiler with a total trim system lag-time of 30 seconds and 30 fire-rate movements an hour, and 2 trims for each movement, that the total time the boiler is not trimmed to setpoint will be 30 x 30 x 2 = 1800 seconds or 30 minutes in each hour; or to put it another way, the conventional trim control will hold the boiler at the precise oxygen level required for maybe 50% of its operating time. Figure 17-2 shows graphically the profound effect of the lag-time problem against the frequency of the fire-rate movements.

It can clearly be seen from Figure 17-2 that it is of vital importance to take into account the number of fire-rate movements before calculating the expected payback from an oxygen trim control, and having established the total lag-time for a specific boiler, only then can you approach reality in terms of estimated fuel savings. As an example, if you have a boiler with 30 movements per hour, a lag-time of 20 seconds, and a proposed oxygen trim system with a sensor in-situ response time of 10 seconds, then your total lag-time is going to be 30 seconds, times an average of 2 trim lag-times required, times 30 fire-rate movements, then the total operating time when the boiler will not be at optimum oxygen setpoint will be, for example:

$$\frac{20 \text{ s.LT} + 10 \text{ s.RT} \times 2 \times 30 \text{ movements}}{3600 \text{ seconds}} = \frac{1800}{3600} = \frac{1800 \times 100}{3600 \cdot 3600} = 50\%$$

Now when you meet the salesman who is trying to sell an oxygen trim system, you can more realistically determine savings and payback. If he says that his control will save $34,000 in a year, you can see from the figures above that in reality it will only be about 50% or $17,000.

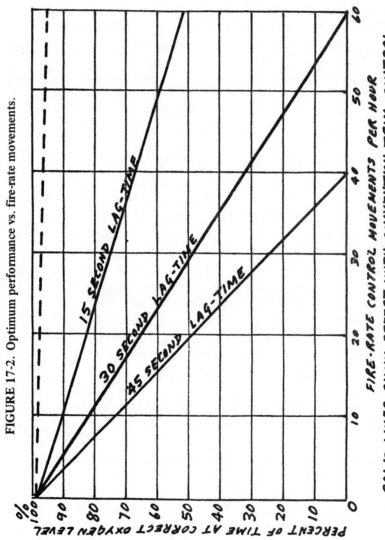

FIGURE 17-2. Optimum performance vs. fire-rate movements.

The reason the discrepancy occurs is because most savings estimates are based on a constant fire-rate or constant load, which never occurs in real boiler operations, and fire-rate movements are not taken into account by most suppliers for obvious reasons. It is worthwhile discussing this issue with possible vendors, if only to determine their depth of understanding.

The proven solution: If you have established that the conventional trim control will give you optimum setpoint performance for less than 90% of your operating time, then your savings can be increased and your boiler controlled to optimum for 95 to 99% of the time by specifying a true Adaptive Trim Control that automatically self-learns and feeds forward actual damper profile position for all fire-rates, without any manual or operator intervention whatsoever.

LEARNING TO ADAPT

The main causes of error in combustion control are: hysteresis (dead-band) and sloppiness through the mechanical profiler and linkages; worn, damaged or dirty burner; sooting up of the boiler tubes; change in viscosity, temperature, pressure, or calorific value of incoming fuel; warm-up effects in the burner and boiler.

The addition of a trim adjustment to air-damper position (or to fuel valve) is normally the preferred means to effect combustion control in oil- or gas-fired boilers where a mechanical air/fuel ratio profiler exists: the trim-only action allows easy reversion to manual control, and because it has limited adjustment, allows the safety and security of the mechanical profiler to be largely retained.

The mechanical profiler is set up during commissioning to give a margin of excess air in order to accommodate the expected changes in burn conditions, and the limitations of the mechanical profiler. The normal action of the trim control is thus to slightly close the air damper in order to achieve the optimal oxygen-control level: trim position will be required to vary as a function of changing fire rate.

A conventional oxygen-control loop (Figure 17-3) controls trim position by feedback from the error between oxygen desired-

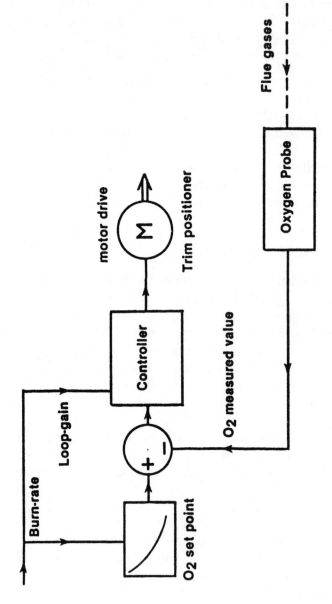

FIGURE 17-3.

value (setpoint) and oxygen measured-value, the oxygen setpoint being profiled to be fire-rate dependent. Because changes in trim position must be provided as fire rate alters, therefore the error-actuated oxygen-control loop must be transiently in error to provide the necessary trim movement.

Peak magnitude of this transient can be severe; the duration with the feedback loop optimally tuned will be in the order of twice the net response-delay of the oxygen-probe plus burner-to-probe lag-time. Typically the transient may be several percent O_2 in magnitude, with duration between 15 seconds for a fast-responding probe or more than a minute with a slow-responding probe.

With the adaptive control system it is possible to achieve immediate oxygen-trim response against varying fire rate, with no oxygen-error transient, so that probe response time becomes of secondary importance. The adaptive system (shown in simplified form in Figure 17-4) utilizes the same basic feedback-loop as does the conventional control of Figure 17-1, with loop-gain compensation from fire rate, and is thus as tightly tuned as possible under all firing conditions.

The controller output corresponding to each narrow range of fire rate is retained in a number of (digital) stores, in which a "trim profile" is thus built up as adaption proceeds with operation over the full range of fire rate. Once adapted, the trim setpoint requirement for each given fire rate is selected immediately, so that no oxygen-error transient is required or then exists.

The effectiveness of the adaptive system as described so far is limited only by hysteresis (dead-band) present in the mechanical profiler and linkages. A transient error can occur, of magnitude dependent upon the extent of the hysteritic dead-band and with duration the same as for the conventional loop of Figure 17-3. The adaptive loop does not result in a worsening of the basic error-loop response time. The hysteresis effect is readily eliminated in a simple restructuring of the adaptive-trim technique, involving measurement of both fuel and damper position to create in effect an adaptive air/fuel profile rather than the adaptive trim/air (or trim/fuel) profile.

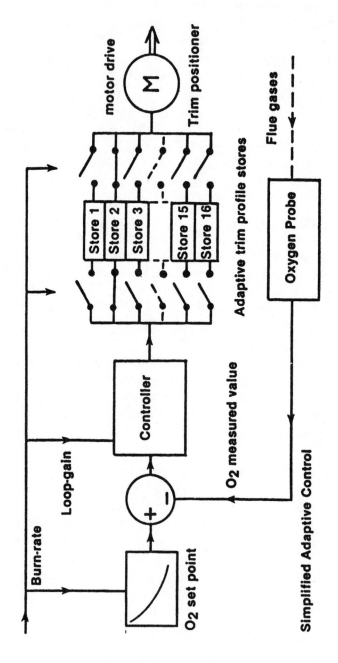

FIGURE 17-4.

Remaining sources of error are mostly slow-varying, and can be followed by self-learning change in the adaptive-profile. To achieve control only requires about 2K-byte extra storage in the processor PROM, and this adds very little to the cost. The fast response time allows operation closer to the ideal oxygen profile, cutting fuel consumption significantly more than can be achieved by conventional trim systems.

A final point of caution: In today's fast changing hi-tech world there are many new terms and buzz words in use. Due to ignorance and sometimes for other reasons, these new words are used incorrectly and it is easy for a potential customer to be confused or misled into thinking he is getting something he really is not. In the field of oxygen trim control, statements like "feed-forward," "learning," and "adaptive" are being misused by some, and can cause misinterpretation by others.

Here is a dictionary definition of adaptive: adaptive/a-'dap-tiv 1: the act or process of adapting; 2: having intelligence to sense conditions of environmental or other intensities or qualities or stimulations; 3: to determine actions or modifications of an organism or parts that make it more fit for existence under current conditions; 4: learning to adapt; 5: something that is adapted becomes more fitted to its function or current situation; 6: adapted to optimum situation; 7: to be in line with current requirements.

According to our research, only three manufacturers in the world currently offer true adaptive trim control. Remember, all trim controls were not created equal and you will in general only get what you pay for. The bottom line is not "how much" but what will the savings and payback really be. If in doubt you should consult an expert!

You should not delay your trim control project because of fears or uncertainties, for two reasons:
1) Every day you delay is wasting more of our precious energy.
2) The new technologies are now well proven; true adaptive oxygen trim is available now, and has been in use by some for more than 3 years with outstanding results.

Conventional O_2 trim controls have been lacking in ability to provide sufficient meaningful data. With the breakthrough in micro-technology, the more advanced true adaptive O_2 trim controls are able to provide a wide range of meaningful data that is displayed and can also be fed to recorders, data-loggers, energy management systems.

To be able to determine the effectiveness of performance most or all of the following data should be available from the control system:

Oxygen Setpoint (fire-rate dependant)
Oxygen
Oxygen Error
Excess Air
Latent Heat Loss
Radiated Heat Loss
Sensible Heat Loss
Stack Loss
Efficiency (thermal – including all losses)
Fire Rate
Rate of change of Cell Temperature

Cell Setpoint Temperature
Cell Temperature
Flue Temperature
Air Temperature
Loop open during Cold Start sequence (= 1)
Lithium Battery Voltage (data storage back-up)
Drive to Neutral Logic (1 = on, 0 = off)
Normal Fire Logic   (1 = on, 0 = off)
Gas Fuel Logic   (1 = on, 0 = off)
Auxiliary Fuel Logic   (1 = on, 0 = off)

Trim Velocity Drive signal
Trim Position (Angular Displacement)
Damper Position Setpoint
Damper Position Measured Value
Damper Position Error
Fuel Valve Position Measured Value
Dry Carbon Dioxide
Adaptive Integrators *1* to *8*
Adaptive Integrators *9* to *16*
Burner Log-Run Time (totalized burner hours)
Burner Log-Start Date

18

Should You Purchase A New Boiler?

As previously mentioned, the cost of adding auxiliary equipment to improve unit performance on an existing boiler may justify the purchase of an entirely new system. This is especially true of an older unit with a limited expected useful lifetime or a unit with severe space or operating constraints that require additional expensive modifications. Other factors may prohibit the use of various equipment or techniques and limit the boiler's efficiency improvement potential.

Once the purchase of a new system has shown to be cost effective, the selection of the most efficient unit incorporating the optimum use of the auxiliary equipment discussed in the previous sections becomes all important. Several manufacturers representing various boiler configurations should be consulted to select the best system on the basis of initial, operating, and maintenance costs. It is generally advised to select standard units available commercially as special design specifications will add significant costs to the purchase of the unit.

When purchasing a new boiler it is very advisable to compare boiler efficiencies of prospective units since even a few tenths percent efficiency advantage can lead to substantial savings in fuel over the life of the boiler. It is common for boiler manufacturers to offer a performance guarantee which will contain at least the guaranteed minimum boiler efficiency at maximum boiler capacity. Predicted performance at intermediate loads may also be provided, especially for larger boilers.

Ideally, the complete efficiency versus load characteristics of the boiler would be available to enable efficiency comparisons at the particular range of loads anticipated for the boiler. Unfortunately, this information is very seldom available and the efficiency characteristics of the boiler are determined only after installation and start-up.

The excess air characteristics of the burner (i.e., excess O_2 versus load profile) are very important in determining the efficiency vs. load curve of the boiler. This is especially true for boilers without waste heat recovery where stack temperatures are higher and the efficiency is consequently more sensitive to the quantity of excess air used. Therefore it may be desirable to avoid burner or furnace designs which require high excess air.

Increased excess air requirements at lower loads is quite common but some combustion control systems and burner designs are available which permit very low excess air levels over the entire turndown range of the boiler. Choosing this type of system would be very attractive especially when the boiler will be operated at lower loads a large fraction of the time.

Other factors not directly connected to boiler costs and efficiency must be considered. Changes in steam requirements or operating conditions may warrant a completely different type, size or firing mode than previously employed in the existing system. Space limitations or steam demand flexibility can make the use of several smaller units more desirable than one large unit.

Consideration should also be given to the "quick steaming" boilers that can come to operating temperature and pressure in a few minutes. These units are very compact based on a once-through design that eliminates the large water jackets of comparable capacity firetube designs.

The steady state operating efficiency of these units is comparable to a conventional boiler at rated loads but they can produce significant fuel savings in applications where the boiler is used for only one shift a day. A conventional firetube boiler radiates a considerable quantity of heat as the water jacket cools after shutdown. This heat is wasted and must be replaced when the boiler is refired for the next day of use. Since the once-through units have a very small water volume, they can produce steam in minutes and save the extra fuel a conventional boiler would use for startup.

Using the boiler manufacturer's example of a 10-hour day and a 40-minute startup period at 81% efficiency for the conventional drum boiler, the once-through steam generator would use 17% less fuel. Some quick steaming boilers also have smaller casing surface area (compared to conventional boilers with equivalent capacity) which reduces radiation heat losses and leads to higher boiler efficiencies, especially at reduced firing rates.

Numerous other options are available to the boiler purchaser. New air pollution constraints and changing fuel supplies may alter the fuels employed.This will strongly influence the method of firing and to a certain extent the configuration of the unit. In all cases, a complete engineering and financial evaluation of possible retrofit applications to the present system and the purchase of a new system is justified prior to the selection and installation of a new steam generating unit.

19

Financial Evaluation Procedures

PERFORMANCE DEFICIENCY COSTS

Procedure

The financial benefits obtained with improved boiler efficiency are readily calculated using

$$\text{Fuel \$ Savings} = W_f \times \frac{\Delta \epsilon}{\epsilon_A} \times C$$

Where:
- W_f = fuel usage at lower efficiency
- $\Delta \epsilon$ = efficiency improvement differential
- ϵ = achievable efficiency level
- C = fuel costs

The units selected must be consistent, that is, if W_f is in millions of Btu/years and C is in dollars per million Btu, the fuel savings will be in dollars per year.

Since the efficiency improvement differential and achievable efficiency level will vary with load, this calculation should be conducted at each major operating load condition. The annual fuel dollar savings will then be the sum of each load condition multiplied by the fraction of operating time spent at each load.

Example

Basis: $W_f = 75,000 \times 10^6$ Btu/yr
   $\Delta\epsilon = 1\%$
     $\epsilon = 80\%$
     $C = \$6/10^6$ Btu

Fuel Savings $= \dfrac{(75,000 \times 10^6 \text{ Btu/yr}) (1\%) (\$6/10^6 \text{ Btu})}{(80\%)}$

     $= 5625/\text{year}$

FIRST AND SECOND LEVEL
MEASURES OF PERFORMANCE

Definitions

First Cost (FC) — cost of labor and materials to implement scheme

Annual Operating Costs (AOC) — all costs (if any) associated with maintaining and operating the scheme

Annual Fuel Savings (AFS) — quantity of fuel saved by use of the installed scheme

Projected Fuel Prices (PFP) — estimated average fuel price over the lifetime of the investment

Estimated Lifetime (EL) — length of time the scheme is expected to be in use

Net Annual Savings (S) — annual fuel savings less operating costs

Depreciation Charge (DC) — annual depreciation of the investment. For a straight line depreciation,
$$DC = \frac{FC}{EL}$$

Discount Rate (DR) — best return on investment

Present Worth Factor (PWF) — factor applied to determine present worth of future savings.

Present Value (PV) — worth of future savings for a given discount rate.

First Level Measure of Performance

This is a short analysis used to screen non-profitable energy conservation opportunities and is not intended to be a complete economic evaluation.

- The *Payback Period (PP)* allows a rough estimate of the period required to recoup the investment.

- PP = FC/S

- A payback period less than half the expected life is considered potentially profitable.

- The *Return on Investment (ROI)* accounts for the depletion of the investment over lifetime by providing for a renewal through depreciation charge.

- $ROI = \dfrac{S\text{-}DC}{FC}$

- If the ROI is less than 20%, a second order level of analysis is required.

Second Level Measure of Performance

These analyses incorporate an allowance for the time value of money, generally as a discount factor.

- The discount is the best available rate of return for the investment dollars which will vary between industries but will usually run 10-20%.

- An investment for energy improvements must yield a return greater than the discount rate.

- The Benefit Cost Analysis provides a direct comparison of the present savings (properly discounted and summed over the expected lifetime) with the initial costs.

- The present savings is determined using the tabulated present worth factors given in Table 19-1 times the net annual savings.

- A benefit/cost ratio greater than unity shows a profitable investment.

TABLE 19-1. Present Worth Factors (PWF)

Lifetime (EL)	Discount Rate (D)				
	5%	10%	15%	20%	25%
1	0.952	0.909	0.870	0.833	0.800
2	1.859	1.736	1.626	1.528	1.440
3	2.723	2.487	2.283	2.106	1.952
4	3.546	3.170	2.855	2.589	2.362
5	4.329	3.791	3.352	2.991	2.689
6	5.076	4.355	3.784	3.326	2.951
7	5.786	4.868	4.160	3.605	3.161
8	6.463	5.335	4.487	3.837	3.329
9	7.108	5.759	4.772	4.031	3.463
10	7.722	6.145	5.019	4.192	3.571
11	8.306	6.495	5.234	4.327	3.656
12	8.863	6.814	5.421	4.439	3.725
13	9.394	7.103	5.583	4.533	3.780
14	9.899	7.367	5.724	4.611	3.824
15	10.380	7.606	5.847	4.675	3.859
16	10.838	7.824	5.954	4.730	3.887
17	11.274	8.022	6.047	4.775	3.910
18	11.690	8.201	6.128	4.812	3.928
19	12.085	8.365	6.198	4.843	3.942
20	12.462	8.514	6.259	4.870	3.954
21	12.821	8.649	6.312	4.891	3.963
22	13.163	8.772	6.359	4.909	3.970
23	13.489	8.883	6.399	4.925	3.976
24	13.799	8.985	6.434	4.937	3.981
25	14.094	9.077	6.464	4.948	3.985

The above table is calculated from the following equation:

$$PWF = \frac{1 - (1 + D)^{-EL}}{D}$$

where D is discount rate expressed as a fraction and EL is the expected lifetime of the project in years.

- This procedure is considered to be superior to all other measures of profitability.

- The *Time to Recoup Investment* is similar to payback period but takes into account the discount rate.

- Locate the payback period in the tabulated values of present worth factors (Table 19-1) for the appropriate discount rate and read the break-even period from the estimated life column.

Example of First and Second Level Measures of Performance

Basis:

- First Cost (FC)   = $400,000
- Annual Fuel Savings (AFS)   = 27,500 MBtu/hr
- Projected Fuel Price (PFP)   = $4.00/MBtu
- Net Annual Savings (S)   = (AFS x PFP) - AOC
                                   = 27,500 MBtu/hr
                                     x 4.00 $/MBtu - 0
                                   = 110,000 per year

- Discount Rate (D)   = 20%
- Expected Lifetime (EL)   = 10 years
- Present Worth Factor (PWF)   = 4.192 (Table 18-1)

- Payback period (No discounting)

$$PP = \frac{FC}{S} = \frac{\$400,000}{\$110,000} = 3.6 \text{ yrs}$$

- Return on Investment

$$DC = \frac{FC}{EL} = \frac{\$400,000}{10 \text{ yr}} = 40,000 \text{ per yr}$$

$$ROI, \frac{\%}{Yr} = \frac{S - DC}{FC} \times 100\% = \frac{(\$110,000 \text{ \$/yr} - \$40,000 \text{ \$/yr})}{\$400,000}$$

$$\times 100\% = 17.5\% \text{ per year}$$

- Benefit/Cost Analysis

PV = S x PWF = $110,000 x 4.192 = $461,120

$$B/C = \frac{PV}{PC} = \frac{\$461,120}{\$400,000} \text{ or } 1.15$$

Time to recoup investment can be quickly approximated by using Table 19-1 and the payback period (PP) estimated earlier as 3.6 years. In the 20% discount rate column one can find that the present worth factor closest to 3.6 is 3.605 which indicates that the investment will be entirely recouped in about 7 years when taking the time value of money into consideration.

MARGINAL ANALYSIS

Marginal analysis is useful in determining the extent of investment for opportunities whose rates of return decrease as the level of investment increases, i.e., to determine the optimum level of application (for example, the application of insulation).

Definition

- Total Costs (TC) — costs over lifetime
- Total Savings (TS) — savings over lifetime
- Marginal Savings (MS) — present value savings generated by last increment of the project
- Marginal Cost (MC) — present value cost of the last increment of the project

Example *(Referring to Figure 19-1)*

- All investments up to Q_1 are profitable since (TS>TC).
- Most profitable level is at Q_0, where the distance between TS and TC is maximized. As illustrated, this corresponds to MS = MC.
- This can also be used to indicate a reducing level of investment for a proposed investment that would first appear unprofitable.

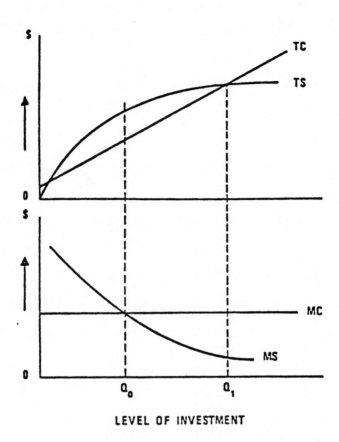

FIGURE 19-1. Determination of optimal investment size based on the marginal savings – marginal cost relationship.

20

Case Studies

NATURAL GAS FUEL

Boiler No. 1

13,000 lbs/hr, D-type water tube boiler with automatic parallel position combustion controls using a single-drive mechanical jackshaft system.

Figure 20-1 shows the dependence of stack parameters on excess O_2 at 80% capacity. Note that CO emissions increase while stack temperatures decrease continuously as excess O_2 is decreased.

Boiler efficiency increases to a maximum, then drops off as CO emissions become excessive (>2000 ppm). NO emissions remain relatively constant during the O_2 variation.

This same dependence is illustrated in Figure 20-2 for operation at 20% of capacity. Overall boiler efficiency decreases due to the radiation losses at this load being more significant even though stack gas temperatures have decreased.

The importance of burner design parameters on these variables at 80% capacity is illustrated in Figure 20-3. The combustion air swirl (effectively the fuel/air mixing and resulting flame patterns) was altered by manually adjusting the burner air register position. Increased air swirl with the air registers closed down one notch from normal setting resulted in a "steep" CO dependence on excess O_2. (A further closing caused excessive flame impingement.) Higher CO emissions resulted from opening the air register position and decreasing the air swirl.

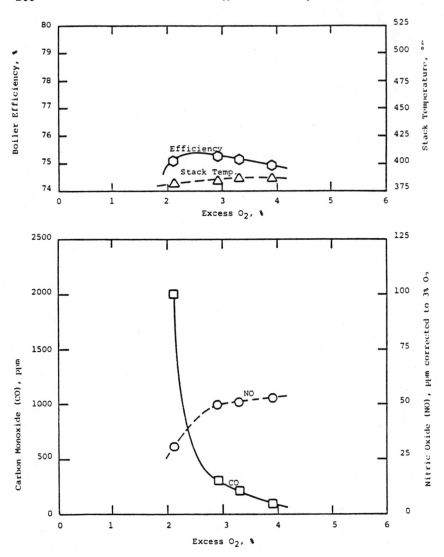

FIGURE 20-1. Boiler No. 1, Natural Gas — Boiler efficiency and stack condition versus excess oxygen at 80 percent load.

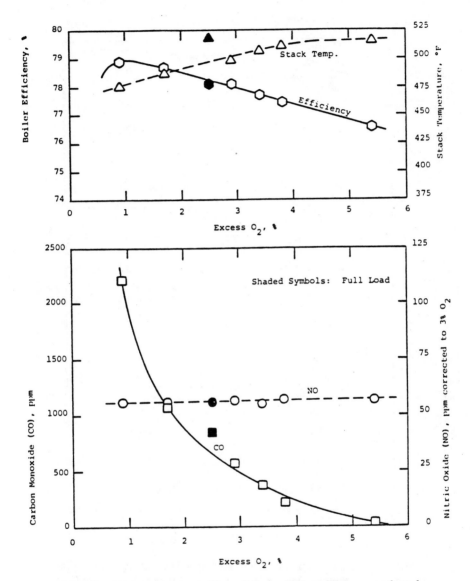

FIGURE 20-2. Boiler No. 1, Natural Gas — Boiler efficiency and stack conditions versus excess oxygen at 20 percent load.

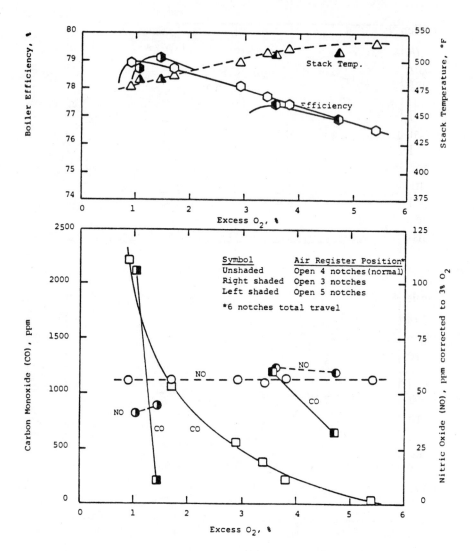

FIGURE 20-3. Boiler No. 1, Natural Gas — The effects of air register position on boiler efficiency and stack conditions at 80 percent load.

Boiler No. 2

15,000 lbs/hr, D-type watertube boiler with automatic parallel positioning combustion controls using a single-drive mechanical jackshaft system.

The dependence of stack parameters on excess O_2 level is illustrated in Figure 20-4.

The "as found" efficiency tests at 60% load indicated very poor combustion efficiency from the presence of up to 50,000 ppm CO and 10,000 ppm unburned hydrocarbon (as CH_4).

The excessively low excess O_2 level at this load was ultimately traced to a loose adjustment screw on the air/fuel control cam. This problem was only encountered at this load.

Boiler No. 3

42,000 lbs/hr watertube boiler with automatic parallel positioning combustion control using a single-driven mechanical jackshaft system.

The "as found" efficiency results for automatic operation are plotted versus load in Figure 20-5 for the entire turndown range of the boiler.

These reveal a rapid decline in boiler efficiency at reduced loads due to exceptionally high excess O_2 levels.

Efficiency also declines at full load due to increased excess O_2 levels caused by restricted natural gas flow.

To determine potential efficiency improvement with lower excess O_2, the air flow was manually reduced while maintaining constant firing rate near 90% capacity as shown in Figure 20-6. A 1% improvement was achieved although the optimum O_2 level of 5.0 was high for natural gas combustion.

Burner air register adjustments to reduce CO could not be attempted.

It was difficult to define a normal operating excess O_2 point for each load due to the considerable play in the control shafts and linkages and therefore the excess O_2 was very non-repeatable.

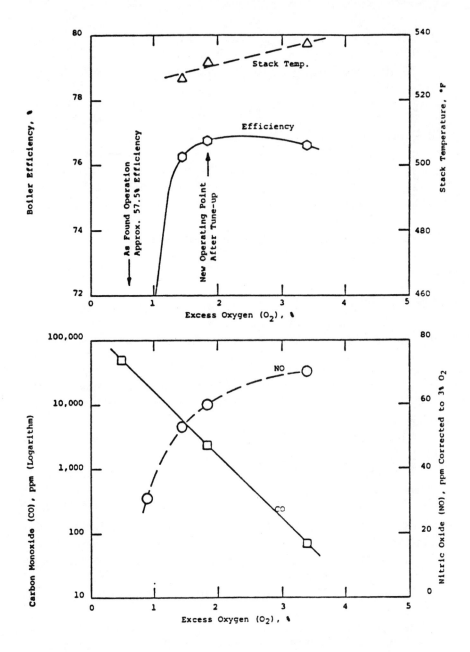

FIGURE 20-4. Boiler No. 2, Natural Gas — Boiler efficiency and stack conditions versus excess O_2 at 65 percent load.

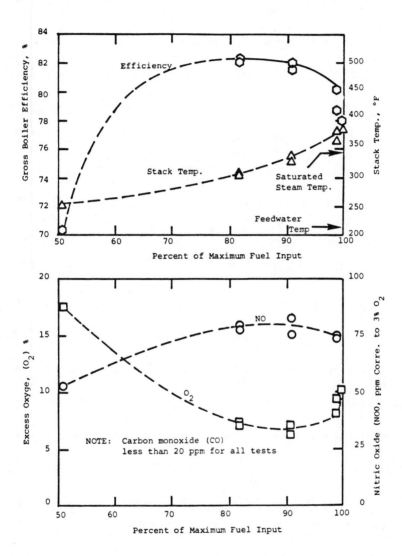

FIGURE 20-5. Boiler No. 3, Natural Gas — Boiler efficiency and stack conditions versus percent of maximum load. (1250 hp, CE). KVB®

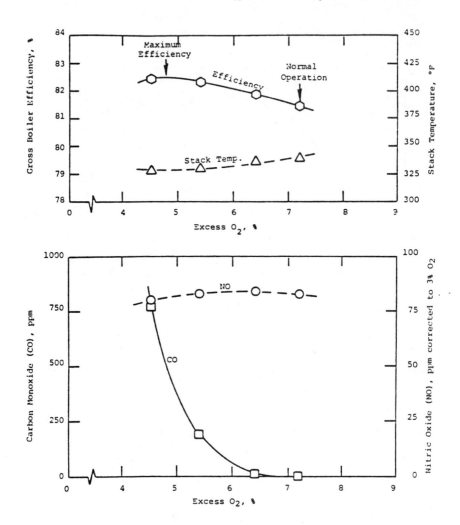

FIGURE 20-6. Boiler No. 3, Natural Gas — Boiler efficiency and stack conditions versus excess oxygen at 90 percent boiler load (1250 hp, CE).

OIL FUEL

Boiler No. 4, No. 2 Diesel Fuel

B&W saturated steam boiler, capacity 58,000 lbs/hr., 230 psig. Single circular burner; Bailey automatic combustion controls (parallel positioning).

Figure 20-7 shows the dependence of efficiency and stack emissions on excess O_2 at constant load.

The lower excess O_2 level (4%) was close to the smoke threshold and probably represents the lowest practical minimum excess O_2 level.

Boiler efficiency was improved by over 1.5% from the normal operating point.

Boiler No. 5, Diesel Fuel

13,000 lbs/hr, D-type watertube boiler with automatic parallel positioning combustion control using a single-drive mechanical jackshaft system.

Figure 20-8 illustrates the dependence of efficiency and stack emissions on excess O_2 over the normal operating load range of the unit.

PULVERIZED COAL

Boiler No. 6

160,000 lbs/hr with four B&W burners arranged in a two-by-two matrix burning Colorado subbituminous coal.

Figure 20-9 illustrates the effects of varied excess O_2 level on stack gas parameters and operating efficiency. As shown, stack gas temperatures were relatively unaffected by excess O_2 variations. Operating efficiency increased by 0.5% with a 1.5% reduction in excess O_2.

Stack gas parameters and efficiency were not significantly affected by variations in operating load as shown in Figure 20-10.

Excess O_2 levels increased with increasing load due to the CO emissions profiles given in Figure 20-11

Unburned CO and carbon in the ash did not play a significant part in efficiency losses.

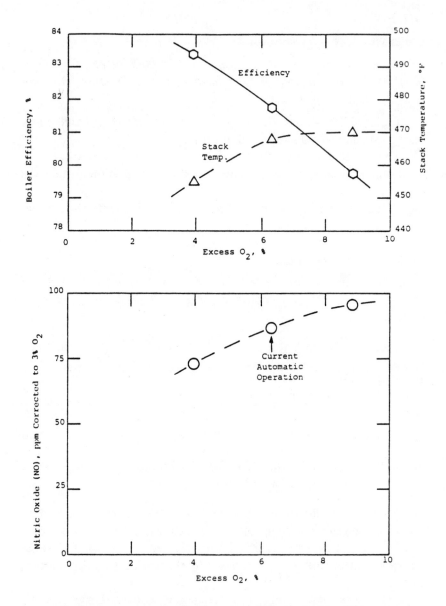

FIGURE 20-7. Boiler No. 4, No. 2 Diesel Fuel – Boiler efficiency and stack conditions versus excess oxygen at half load (24 K lb/hr).

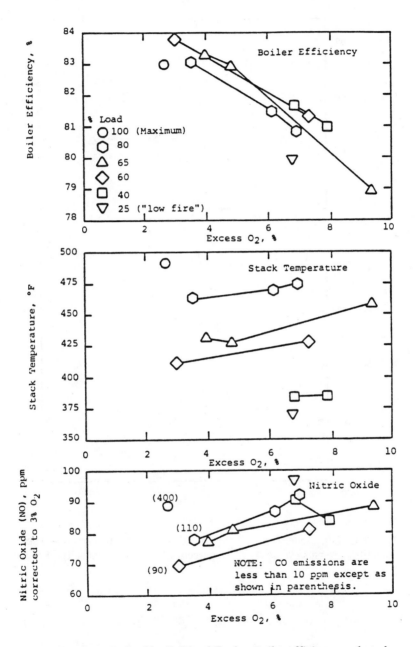

FIGURE 20-8. Boiler No. 5, Diesel Fuel — Boiler efficiency and stack conditions versus excess oxygen at various boiler loads.

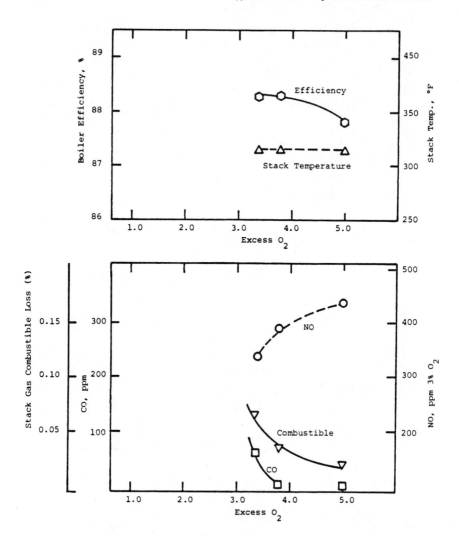

FIGURE 20-9. Boiler No. 6, Pulverized Coal — Boiler efficiency and stack conditions versus excess O_2 at 70 percent of maximum capacity. KVB®

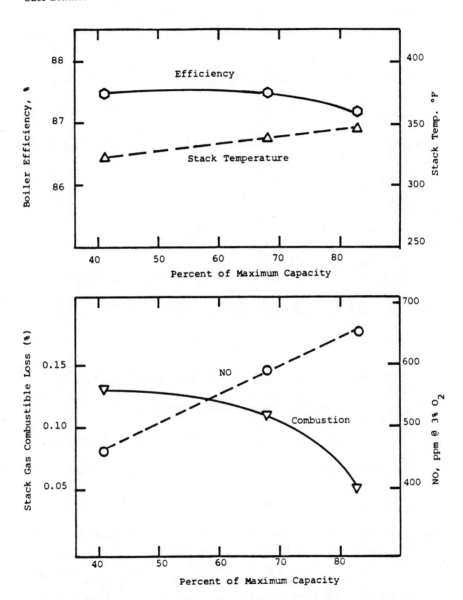

FIGURE 20-10. Boiler No. 6, Pulverized Coal — Boiler efficiency and stack conditions versus percent of maximum capacity. KVB®

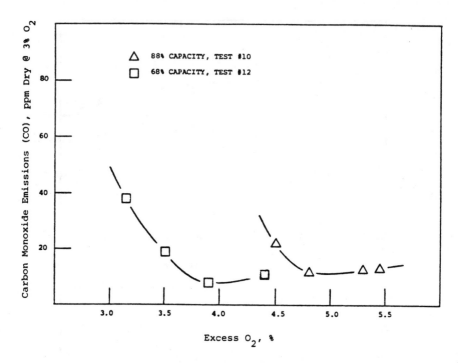

FIGURE 20-11. Boiler No. 6, Pulverized Coal — Carbon monoxide emissions as a function of excess oxygen at three boiler loads, Walden Colorado coal, Fremont Department of Utilities.

Boiler No. 7

225,000 lbs/hr, corner-fired watertube unit fired with eight pulverized coal burners burning high sulfur bituminous coal.

At constant load conditions, variations in excess air had little effect on stack gas temperatures as shown in Figure 20-12. A 0.5 percent increase in efficiency with a 1.5% drop in excess O_2.

NO emissions increased with increasing excess O_2.

Efficiency decreased with increased load due to increased stack gas temperatures, illustrated in Figure 20-13.

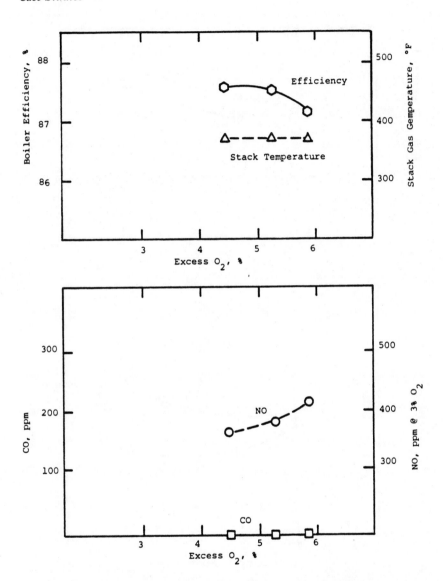

FIGURE 20-12. Boiler No. 7, Pulverized Coal — Boiler efficiency and stack conditions versus excess O_2 at 100 percent capacity. KVB®

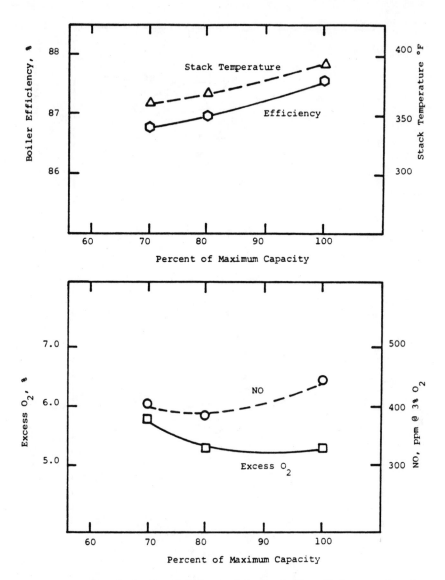

FIGURE 20-13. Boiler No. 7, Pulverized Coal — Boiler efficiency and stack
conditions versus percent of maximum capacity. KVB®

STOKER-FIRED COAL

Boiler No. 8

100,000 lbs/hr Westinghouse centerfire spreader stoker with traveling grate firing high sulfur bituminous coal.

The relationship of stack gas parameters with excess O_2 is presented in Figure 20-14. CO and combustible losses increase sharply as excess O_2 is lowered below 7.5% O_2. Efficiency, however, has not yet peaked since the stack gas temperatures are decreasing.

As given in Figure 20-15, efficiency also decreases with decreasing load due to increase in excess O_2 levels and combustible emissions.

Boiler No. 9

160,000 lbs/hr watertube B&W boiler fired by six spreader stokers supplied by the Detroit Stoker Company. Low sulfur subbituminous coal.

As illustrated in Figure 20-16, stack gas parameters and boiler efficiency are significantly affected by variations in excess O_2. Note the sharp increase of CO and combustible emissions and decrease in exit gas temperature as excess O_2 is lowered below 7.0% O_2.

An analysis of these parameters as load is varied (Figure 20-17) would indicate that the operating efficiency over the load range would be relatively constant if the excess O_2 were increased at higher loads to reduce CO and combustible emissions.

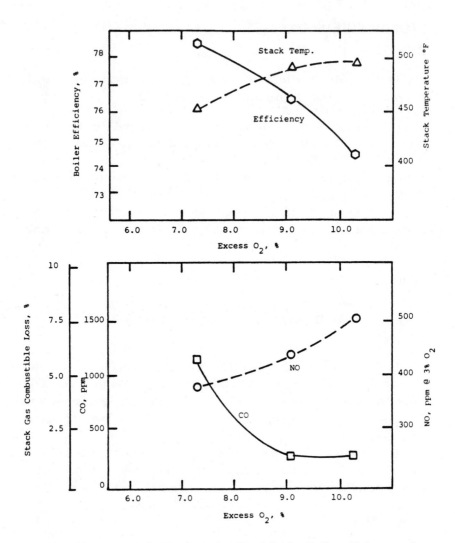

FIGURE 20-14. Boiler No. 8, Stoker-Fired Coal — Boiler efficiency and stack conditions versus excess at 98 percent maximum capacity. KVB®

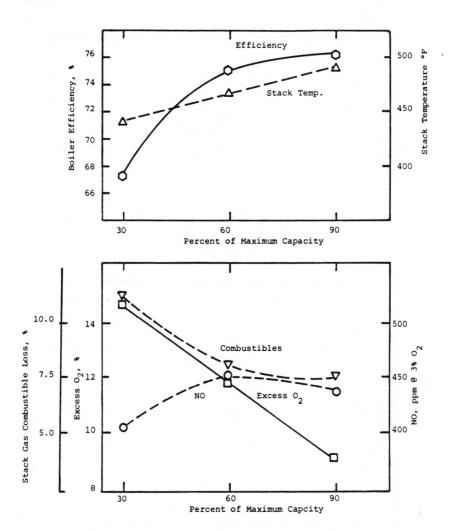

FIGURE 20-15. Boiler No. 8, Stoker-Fired Coal — Boiler efficiency and stack condition versus percent maximum capacity. KVB®

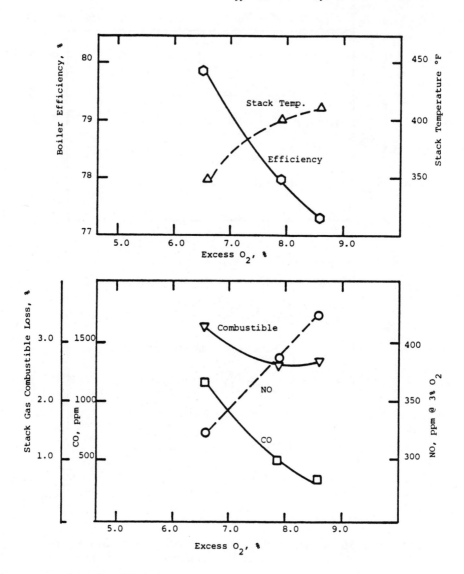

FIGURE 20-16. Boiler No. 9, Stoker-Fired Coal — Boiler efficiency and stack condition versus excess O_2 at 60 percent capacity.

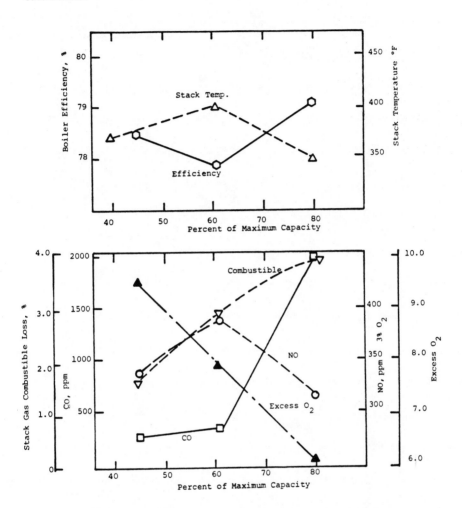

FIGURE 20-17. Boiler No. 9, Stoker-Fired Coal — Boiler efficiency and stack conditions versus percent of maximum capacity. KVB®

Appendix A

Combustion-Generated Air Pollutants

The exhaust gases from all combustion devices contain a variety of byproducts, some of which are considered environmental pollutants. Some of these pollutants are immediately toxic in themselves while others pose indirect health hazards by virtue of their ability to react with other pollutants in the air to form more hazardous compounds. Furthermore, these pollutants also differ by the way they originate in the combustion processes. Some are more related to the composition of the fuel itself whereas others are dependent on the characteristics of the actual fuel-burning process and as such are sensitive to the design and operating variables of the combustion device.

The United States Environmental Protection Agency (EPA) has presently identified six major pollutants and has established emission standards to protect the public health and welfare. These six "criteria" pollutants are:

1. Particulate matter
2. Sulfur dioxide
3. Nitrogen oxides
4. Carbon monoxide
5. Hydrocarbons
6. Oxidants

Below are brief descriptions of these substances including the principal health effects, their formation in boiler combustion systems and methods currently used to limit their emissions from boiler stacks. This information is included in the manual to familiarize boiler operators and others with the major aspects of combustion pollutants which have prompted action by the EPA and some state and local air quality agencies to limit these emissions from many sources. Frequently, emission regulations or pollutant-related operating criteria are imposed but the operators of the combustion devices are not acquainted with the underlying health and environmental concerns.

Particulate Matter

Particulate matter is the non-gaseous portion of the combustion exhaust, consisting of all solid and liquid material (except water droplets) suspended in the exhaust gases. They can be generally defined as any material that would not pass through a very fine filter. Particulates originating in the combustion process can range in size from submicron in diameter (less than one millionth of a meter) to diameters larger than a millimeter (thousandth of a meter). The larger particulates do not carry far in the atmosphere and usually fall to the ground near the source. The small particles, which may make up the bulk of particulate matter, can remain in the atmosphere for long periods of time and contribute to haze and obscured long-range visibility. These "fine particles" are potentially the most hazardous to health since they are easily carried into the small passages of the respiratory tract during normal breathing.

Particulate matter can be composed of a wide variety of materials including unburned fuel, sulfur compounds, carbon, ash constituents in the fuel (including many toxic metals), and even non-combustible airborne dust that enters the combustion system with the combustion air. Many of these materials by themselves are indentified as health hazards and when dispersed in the air in fine particles, can be inhaled and subsequently absorbed into the body. High concentrations of particulate matter in the inhaled air can have more direct health effects by irritating or blocking the surfaces of the respiratory tract leading to temporary or permanent breathing impairments and physical damage. Oral ingestion of particulate matter is also possible from particulate fall-out onto vegetation and food crops.

The quantity and composition of particulates generated in boilers are influenced by several factors including the type of fuel being burned, the boiler operating mode, and combustion characteristics of the burners and furnace. Burner designs and operating modes that tend to promote thorough, efficient combustion generally will reduce the fraction of combustible material in the particulates. From a boiler operating standpoint, it is desired to minimize these fuel-derived materials since they represent wasted available fuel energy and can lead to troublesome internal furnace deposits or objectionable smoke at the stack. Burner and/fuel ratio is one of the more important operating parameters that can influence the quantity of combustible particulate material generated.

Natural gas and most oil fuels are referred to as "clean burning" fuels primarily due to their lower tendency to form solid combustibles (smoke, soot, carbon, etc.) and their low ash content. By comparison, some heavy oils and most coal fuels contain substantial quantities of ash which subsequently form the bulk of the non-combustible particulate matter generated in the furnace. For the case of coal, ash can compose up to 20 percent or more of the total weight of fuel burned. Preventing its accumulation on internal boiler surfaces is a major consideration in the design of the boiler.

In some boiler designs, most of the coal ash remains in the exhaust gases leaving the boiler, but the use of various particulate control devices allows low concentrations to be emitted from the stack. Current devices employ various techniques to remove particulates from the stack gases. These include filtration, mechanical separation and electrostatic precipitation. Many designs have proven to be capable of removing more than 99% of the particulates and are also applicable to oil-fired boilers where particulate controls are required.

Sulfur Dioxide

SO_2 is a non-flammable, colorless gas that can be "tasted" in concentrations of less than 1 part per million in the air. In higher concentrations, it has a pungent, irritating odor.

Sulfur dioxide (chemical symbol, SO_2) is formed during the combustion process when sulfur (S) contained in the fuel combines with oxygen (O_2) from the combustion air. Sulfur trioxide (SO_3), is another oxide of sulfur which is also formed in this manner. SO_2 together with SO_3 comprise the total oxides of sulfur, generally referred to as "SO_x." SO_3 is usually no more than 3 to 5 percent of the total SO_x generated in the boiler.

Except for sulfur compounds present in particulate matter, all of the sulfur initially contained in the fuel is converted to SO_2

and SO_3. Before leaving the stack, SO_3 can combine with moisture in the exhaust gases to form sulfuric acid which condenses onto particulates or remains suspended in the stack gases in the form of an acid mist. In the atmosphere, a portion of the SO_2 is converted to SO_3 which similarly forms sulfuric acid by combining with moisture in the air. The SO_3 can also form other sulfur compounds such as sulfates.

The sulfates and acid mists can contribute significantly to reduced visibility in the atmosphere. Corrosion of materials exposed to the air and damage to vegetation are other major environmental effects. The health effects of sulfur oxides, sulfuric acids and some of the sulfates are primarily related to irritation of the respiratory tract. These effects may be temporary or permanent and include constriction of lung passages and damage to lung surfaces.

The quantity of SO_x generated in the boiler is primarily dependent on the amount of sulfur in the fuel and is not highly affected by boiler operating conditions or design. Regulating the quantity of sulfur allowed in the fuel is a primary method of controlling SO_x emissions. Stack gas "scrubbers" which remove SO_2 from the combustion exhaust gases can also be effective where "high sulfur" fuels are used.

Nitrogen Oxides

Nitric oxide (NO) and nitrogen dioxide (NO_2) are the two forms of nitrogen oxides generated by boiler combustion processes. Together, these compounds are customarily referred to as total oxides of nitrogen or simply "NO_x." NO is a colorless, odorless gas and is not considered a direct threat to health at concentrations found in the atmosphere. NO_2 is a considerably more harmful substance. Although NO_2 comprises typically 5% or less of the NO_x emitted from boiler stacks, a large fraction of the NO is converted to NO_2 in the atmosphere.

NO_2 is a yellow-brown colored gas which can affect atmospheric visibility. It also has a pungent, sweetish odor that can be detected at concentrations sometimes reached in polluted air. In much higher concentrations (100 ppm) NO_2 can be fatal when inhaled. Prolonged exposures at much lower concentrations can cause cumulative lung damage and respiratory disease.

NO_x is formed spontaneously during the combustion process when oxygen and nitrogen are present at high temperatures. All three ingredients (oxygen, nitrogen and high temperature) are essential elements of the combustion process and it would therefore be very difficult to prevent the formation of NO_x altogether. Nitrogen is present in the combustion air and in the fuel itself. Minimizing the fuel nitrogen content has been shown to reduce NO_x but this is not currently a practical NO_x control approach. Most NO_x reduction techniques currently applied to boilers are effective as a result of lower peak flame temperatures in the furnace, reduced availability of oxygen in the flame or a combination of both. These techniques include low excess air operation, fuel-rich ("staged") firing and flue gas recirculation. While these approaches limit the formation of NO_x in the furnace, future techniques may be developed that "scrub" the NO_x from the exhaust gases before entering the stack. Fuel desulfurization processes and other fuel "cleaning" treatments may have some associated benefits in reduced fuel nitrogen content.

Carbon Monoxide

Carbon monoxide (CO) is a product of incomplete combustion and its concentration in boiler exhaust gas is usually sensitive to boiler operating conditions. For example, improper burner settings, deteriorated burner parts and insufficient air for combustion can lead to high CO emissions. CO measurements at the stack are often used as an indicator of poor combustion conditions.

CO is an invisible, odorless, tasteless gas. Exposure to CO-containing exhaust gases produces a well known "CO poisoning" which can be fatal. CO emitted from boiler stacks are dispersed in the atmosphere and together with CO from other sources are generally not in high enough concentrations to produce any immediate health effects (an exception might be in the vicinity of high density automobile traffic). However, there is concern that long-term exposures to these concentrations may cause eventual health problems.

Hydrocarbons

Like carbon monoxide, hydrocarbons are indicative of incomplete or inefficient combustion and can be essentially eliminated from the boiler stack gases by proper operation of the fuel-burning equipment. However, this may be misleading since, strictly speaking, hydrocarbons cannot be entirely eliminated and trace quantities of hydrocarbon compounds will nearly always be present, regardless of how the boiler is operated.

Due to the tremendous variety of hydrocarbon compounds involved and the unknown health effects of some of these even in much larger quantities, it is difficult to assess their environmental impact. Some of these hydrocarbons resemble actual components in the fuel and are rightfully called "unburned fuel," while others are entirely modified forms generated in complex chemical reactions during the combustion process.

It is known that hydrocarbon air pollutants are important ingredients in the formation of photochemical smog. Under certain atmospheric conditions, they can also be transformed into other derivatives which are potentially more hazardous. Some of the manifestations of smog such as irritation of the eyes and respiratory tract are in part directly associated with hydrocarbons and their derivatives.

Oxidants

The term "oxidant" is generally applied to oxygen-bearing substances that take part in complex chemical reactions in polluted atmospheres. These so-called photochemical reactions, which are often intensified in the presence of sunlight, involve nitrogen oxides and reactive organic substances (including hydrocarbons and their derivatives) as the principal chemical ingredients. These react to form new compounds including ozone and PAN (peroxyacyl nitrates) which are usually considered to be the major oxidants in photo-chemical air pollution ("smog"). The term oxidant is also sometimes used to include the original chemical reactants NO_x, hydrocarbons, and others.

While ozone and PAN are not generated directly in the boiler, the principal ingredients (especially NO_x) are supplied in part by exhaust gases from the boiler. By reducing the emission of these compounds from boilers and all the other combustion sources (automobiles, airplanes, etc.), their photochemical byproducts—the oxidants—will be reduced.

Photochemical oxidants produce adverse effects on vegetable matter which can affect growth, and the quantity and quality of agricultural yields and other plant products. Deterioration of various materials (especially rubber) is also a well known occurrence in polluted atmosphere which is attributed mainly to the presence of ozone.

A major effect on humans is irritation of the eyes. In quantities higher than typically found in polluted atmospheres, oxidants have an irritating effect on the respiratory tract, producing coughing and choking. Headache and severe fatigue may be other side-effects. In lower concentrations found in polluted air, the effects appear less well defined. Aggravation of existing respiratory ailments such as asthma has sometimes been attributed to oxidants.

Appendix B

Conversion Factors

SI Units to Metric or English Units

To Obtain	From	Multiply By	To Obtain ppm at 3% O_2 of	Multiply Concentration in ng/J by
g/Mcal	ng/J	0.004186		
10^6 Btu	GJ	0.948	Natural Gas Fuel	
MBH/ft^2	GJ·hr^{-1}·m^{-2}	0.08806	CO	3.23
MBH/ft^3	GJ·hr^{-1}·m^{-3}	0.02684	HC	5.65
Btu	gm cal	3.9685×10^{-3}	NO or NOx	1.96
10^3 lb/hr* or MBH	GJ/hr	0.948	SO_2 or SOx	1.41
lb/MBtu	ng/J	0.00233		
ft	m	3.281	Oil Fuel	
in	cm	0.3937	CO	2.93
ft^2	m^2	10.764	HC	5.13
ft^3	m^3	35.314	NO or NOx	1.78
lb	Kg	2.205	SO_2 or SOx	1.28
Fahrenheit	Celsius	$t_F = 9/5(t_C) + 32$	Coal Fuel	
	Kelvin	$t_F = 1.8K - 460$		
psig	Pa	$P_{psig} = (P_{pa})(1.450 \times 10^{-4}) - 14.7$	CO	2.69

(continued)

To Obtain	From	Multiply By
psia	Pa	$P_{psia} = (P_{pa})(1.450 \times 10^{-4})$
iwg (39.2°F)	Pa	$P_{iwg} = (P_{pa})(4.014 \times 10^{-3})$

*lb/hr of equivalent saturated steam

To Obtain ppm at 3% O_2 of	Multiply Concentration in ng/J by
HC	4.69
NO or NOx	1.64
SO_2 or SOx	1.18
Refinery Gas Fuel (Location 33)	
CO	3.27
HC	5.71
NO or NOx	1.99
SO_2 or SOx	1.43
Refinery Gas Fuel (Location 39)	
CO	3.25
HC	5.68
NO or NOx	1.98
SO_2 or SOx	1.42

(end)

English and Metric Units to SI Units

To Obtain	From	Multiply By
ng/J	lb/MBtu	430
ng/J	g/Mcal	239
$GJ \cdot hr^{-1} \cdot m^{-2}$	MBH/ft^2	11.356
$GJ \cdot hr^{-1} \cdot m^{-3}$	MBH/ft^3	37.257
GH/hr	10^3 lb/hr* or 10^6 Btu/hr	1.055
m	ft	0.3048
cm	in	2.54
m^2	ft^2	0.0929
m^3	ft^3	0.02832
Kg	lb	0.4536
Celsius	Fahrenheit	$t_c = 5/9 \, (t_F - 32)$
Kelvin		$t_K = 5/9 \, (t_F - 32) + 273$
Pa	psig	$P_{pa} = (P_{psig} + 14.7)(6.895 \times 10^3)$
Pa	psia	$P_{pa} = (P_{psia})(6.895 \times 10^3)$
Pa	iwg (39.2°F)	$P_{pa} = (P_{iwg})(249.1)$

*lb/hr of equivalent saturated steam

To Obtain ng/J of	Multiply Concentration in ppm at 3% O_2 by
Natural Gas Fuel	
CO	0.310
HC	0.177
NO or NOx (as equivalent NO_2)	0.510
SO_2 or SOx	0.709
Oil Fuel	
CO	0.341
HC	0.195
NO or NOx (as equivalent NO_2)	0.561
SO_2 or SOx	0.780
Coal Fuel	
CO	0.372
HC	0.213
NO or NOx (as equivalent NO_2)	0.611
SO_2 or SOx	0.850

(continued)

To Obtain	From	Multiply By

To Obtain ng/J of	Multiply Concentration in ppm at 3% O_2 by
Refinery Gas Fuel (Location 33)	
CO	0.306
HC	0.175
NO or NOx (as equivalent NO_2)	0.503
SO_2 or SOx	0.700
Refinery Gas Fuel (location 39)	
CO	0.308
HC	0.176
NO or NOx (as equivalent NO_2)	0.506
SO_2 or SOx	0.703

(end)

INDEX